M. Albert Vannice

Solutions Manual for

Kinetics of Catalytic Reactions

 Springer

M. Albert Vannice
William H. Joyce Chaired Professor
Department of Chemical Engineering
The Pennsylvania State University
University Park, PA 16802
mavche@engr.psu.edu

ISBN-10: 0-387-25973-2
ISBN-13: 978-0387-25973-4

Printed on acid-free paper.

Printed in the United States of America.   (IBT)

9 8 7 6 5 4 3 2 1

springeronline.com

Solutions Manual for

Kinetics of Catalytic Reactions

Preface

This manual of solutions to the problems in "Kinetics of Catalytic Reactions" has been prepared to assist those who use this book in a teaching function. However, these solutions should also benefit those outside the classroom who want to apply the principles and concepts that are discussed in the book. By studying and observing the approaches used in solving these problems, it is very likely that similar applications can be envisioned in different kinetic problems that the investigator might face. Thus the availability of these solutions is a good learning tool for everyone. Additional details and insight about the solutions provided can be obtained by reading the cited references.

I have tried to eliminate all errors, both conceptual and typographical, in these solutions; however, the probability is high that I have not succeeded completely. Should any errors of commission (or omission) be found, I would greatly appreciate being informed. I can be reached at this email address: mavche@engr.psu.edu, or mail can be sent to me at: 107 Fenske Laboratory, Department of Chemical Engineering, The Pennsylvania State University, University Park, PA 16802.

Albert Vannice

Contents

Problem 3.1 Solution

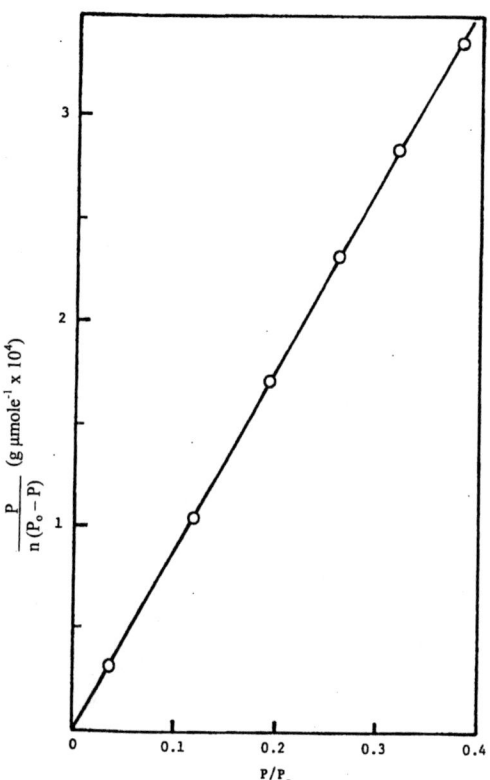

N₂ BET Plot for BC-1

From slope and intercept:
$n_m = 1120$ μmole N_2/g
$C \cong 1000$

$$A_m = \left(\frac{1120 \times 10^{-6}\,\text{mole}}{g}\right)\left(\frac{6.023 \times 10^{23}\,\text{molecule}}{\text{mole}}\right)\left(\frac{16.2\,\mathring{A}^2}{\text{molecule}}\right)\left(\frac{10^{-20}\,m^2}{\mathring{A}^2}\right) = \frac{110\,m^2}{g}$$

$$\ell n\,C = \frac{q_1 - q_L}{RT} \Rightarrow q_1 = 6.9\left(\frac{1.987\,\text{cal}}{\text{mole}\cdot K}\right)(80\,K) + 1340 = \frac{2440\,\text{cal}}{g\,\text{mole}}$$

Problem 3.2(a) Solution

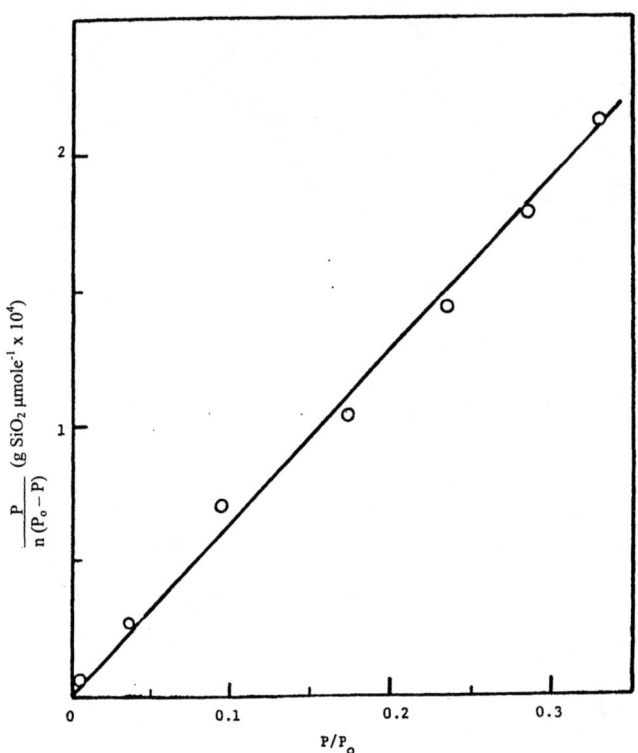

N₂ Bet Plot for SiO₂ (Cab-O-Sil, Grade M5)

From slope and intercept:
$n_m = 1620$ µmole N_2/g
$C = 140$

$$A_m = \left(1620 \frac{\mu mole \ N_2}{g}\right)\left(\frac{6.023 \times 10^{17} \ molecule}{\mu mole}\right)\left(\frac{16.2 \ \mathring{A}^2}{molecule}\right)\left(\frac{10^{-20} \ m^2}{\mathring{A}^2}\right) = \frac{158 \ m^2}{g}$$

$$\ell n \ C = \frac{q_1 - q_L}{RT} = 4.94$$

$$q_1 = 4.94(1.987 \ cal/mole \cdot K)(77 \ K) + 1340 \ cal/mole = \frac{2100 \ cal}{mole}$$

Problem 3.2(b) Solution

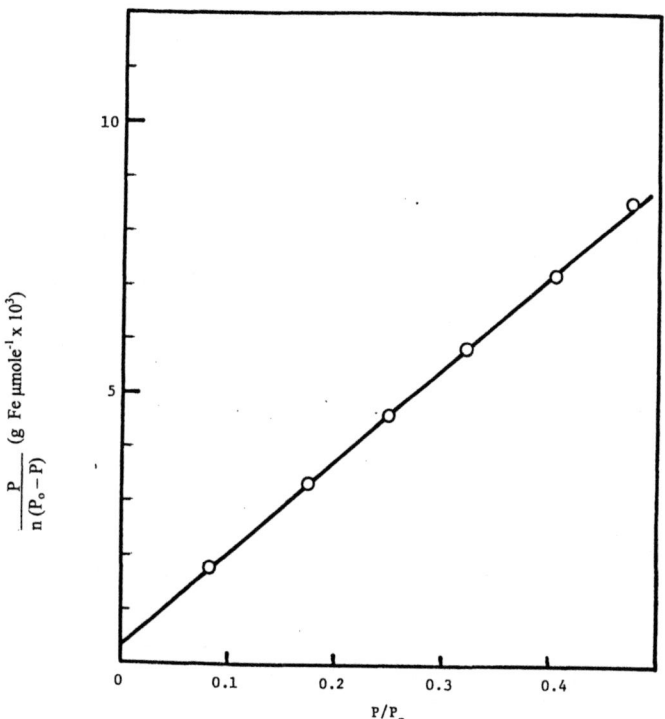

Y-axis: $\dfrac{P}{n(P_o - P)}$ (g Fe μmole^{-1} x 10^3)

X-axis: P/P_o

N$_2$ BET Plot for Fresh, Reduced Fe$_2$O$_3$

From slope and intercept:
$n_m = 10.00$ μmole N$_2$/g
$C = 47$

$$A_m = \left(\frac{10.00 \times 10^6 \text{ mole N}_2}{g}\right)\left(\frac{6.023 \times 10^{23} \text{ molecule}}{\text{mole}}\right)\left(\frac{16.2 \, \overset{\circ}{A}^2}{\text{molecule}}\right)\left(\frac{10^{-20} \text{ m}^2}{\overset{\circ}{A}^2}\right) = \frac{0.98 \text{ m}^2}{g}$$

$$\ell n \, C = \frac{q_1 - q_L}{RT} \Rightarrow q_1 = 3.85$$

$$q_1 = 3.85 \left(1.987 \text{ cal/mole} \cdot K\right)\left(80 \text{ K}\right) + 1340 \text{ cal/mole} = 1950 \text{ cal/mole}$$

Problem 3.2(c) Solution

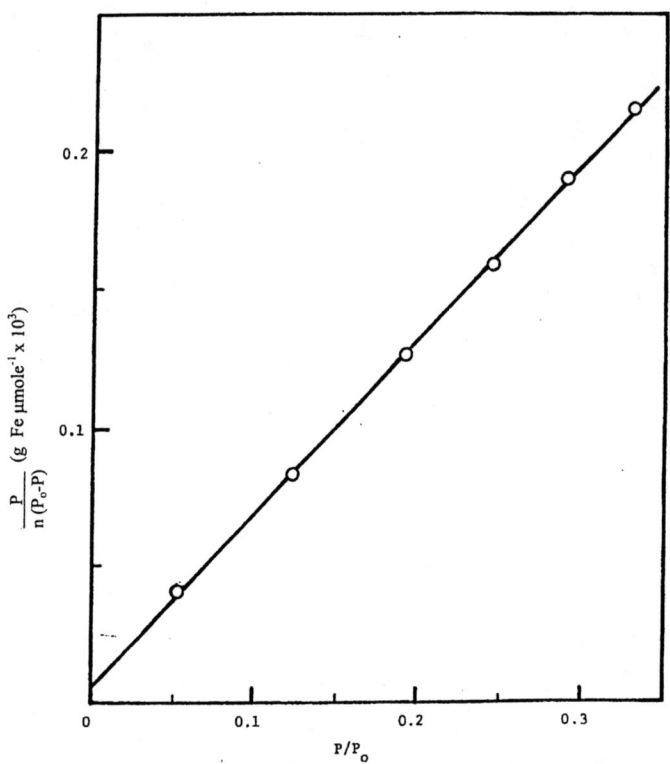

Ar Bet Plot for Used Bulk Iron

From slope and intercept:
$n_m = 1.60$ μmole Ar/g
$C = 47$

$$A_m = \left(\frac{1.60 \times 10^{-6} \text{ mole Ar}}{g}\right)\left(\frac{6.023 \times 10^{23} \text{ molecule}}{\text{mole}}\right)\left(\frac{13.9 \text{ Å}^2}{\text{molecule}}\right)\left(\frac{10^{-20} \text{ m}^2}{\text{Å}^2}\right) = \frac{0.134 \text{ m}^2}{g}$$

$$\ell n\, C = \frac{q_1 - q_L}{RT} = 3.85$$

$$q_1 = 3.85\,(1.987 \text{ cal/mole} \cdot \text{K})(77 \text{ K}) + 1550 = 2.14 \text{ kcal/mole}$$

Problem 3.3 Solution

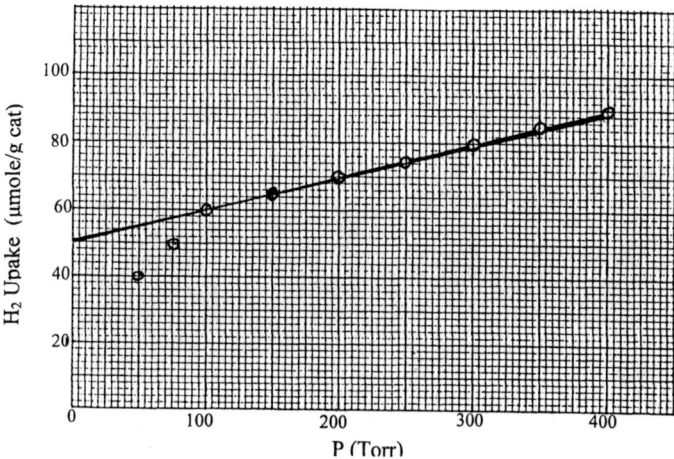

Assume $H_{ad}/Rh_s = 1$. The atomic weight of Rh is 103. The amount of hydrogen chemisorbed on the Rh is the intercept value of 50 μmole H_2/g which is determined by the high-P region where the Rh surface is essentially saturated with H atoms.

$$\frac{\left(\dfrac{50\ \mu\text{mole H}_2}{\text{g cat}}\right)\left(\dfrac{2H_{ad}}{H_2}\right)\left(\dfrac{1\ Rh_s}{1\ H_{ad}}\right)}{\left(\dfrac{0.015\ \text{g Rh}}{\text{g cat}}\right)\left(\dfrac{10^6\ \mu\text{mole Rh}}{103\ \text{g Rh}}\right)} = 0.69 = D_M$$

Problem 3.4 Solution

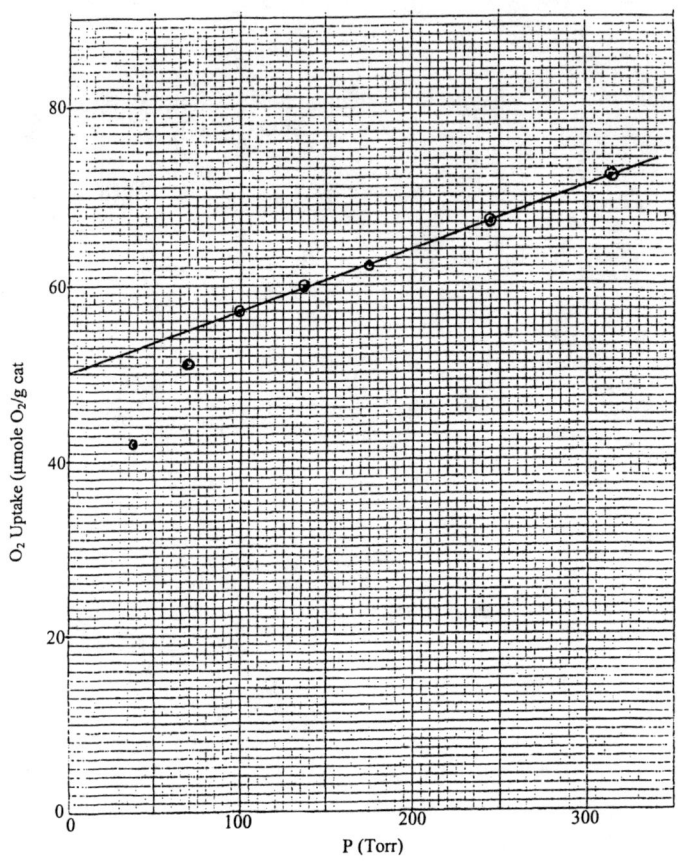

Assume no O_2 adsorption on SiO_2 and the O_{ad}/Ag_s ratio is 1

Adsorption on Ag = 50 μmole O_2/g cat as determined by the intercept determined in the high-P region of saturation.

Dispersion = $Ag_s/Ag_{total} = O_{ad}/Ag_t$

$$D_M = \frac{\left(50 \, \mu mole \, O_2/g\right)\left(2 \, O_{ad}/O_2\right)\left(1 \, Ag_s/O_{ad}\right)}{\left(\dfrac{0.0243 \, g \, Ag}{g \, cat}\right)\left(\dfrac{mole \, Ag}{108 \, g \, Ag}\right)\left(\dfrac{10^6 \, \mu mole}{mole}\right)} = 0.44$$

Problem 4.1 Solution

a) For isothermal, 1st-order reaction:

$$\bar{\eta} = C_s/C_o = 1 - \bar{\eta}Da_o \qquad \text{(Eq. 4.54)}$$

$$\bar{\eta}Da_o = \mathfrak{R}/k_g aC_o \qquad \text{(Eq. 4.50)}$$

$$a = \frac{A}{V} = \frac{2\pi r^2 + 2\pi rL}{L\pi r^2} = 18.75 \text{ cm}^{-1} \quad \text{for cylinder}$$

(For a sphere, $a = \dfrac{A}{V} = \dfrac{6}{2r} = 18.75 \text{ cm}^{-1}$)

$$C_{oSO_2} = \frac{P}{RT} = \frac{0.06 \text{ atm}}{\left(\dfrac{82.06 \text{ atm} \cdot \text{cm}^3}{\text{g mole} \cdot \text{K}}\right)(763.2 \text{ K})} = 9.58 \times 10^{-7} \frac{\text{mole SO}_2}{\text{cm}^3}$$

See Table 1 for $\bar{\eta}Da_o$ and $\bar{\eta}$ values.

b) The concentration drop can be determined from Eq. 4.52, i.e., $C_s/C_o = 1 - \bar{\eta}Da_o$ so $C_o - C_s = \mathfrak{R}/k_g a = \Delta C$

The ΔC values are given in Table 1. The concentration gradients through the film thickness, δ, will be $\Delta C/\delta$.

From Eq. 4.44, $k_g = D/\delta$ so $\delta = D/k_g$.

From Eq. 4.75, the bulk diffusivity is: $D_b = \dfrac{\bar{v}\lambda}{3}$. The mean-free path in the gas phase is (See Illustration 4.6):

$$\lambda = \frac{1}{\sqrt{2}\pi\sigma^2(N/V)} = \frac{RT}{\sqrt{2}\pi\sigma^2 P}$$

for an ideal gas where $\sigma = $ is the molecular cross-section area, so

$$\lambda = \frac{\left(\dfrac{82.06 \text{ atm cm}^3}{\text{g mole} \cdot \text{K}}\right)(763.2 \text{ K})\left(\dfrac{1 \text{ mole}}{6.02 \times 10^{23} \text{ molecule}}\right)}{\sqrt{2}\pi\left[\dfrac{(4 \times 10^{-8} \text{ cm})^2}{\text{molecule}}\right](1 \text{ atm})} = 1.5 \times 10^{-5} \text{ cm}$$

$$\overline{v} = \left(\frac{8k_B T}{\pi m} \right)^{1/2} \qquad \text{(Eq. 4.76)}$$

$$\overline{v} = \left[\frac{8 \left(1.38 \times 10^{-16} \text{ erg/K} \right) \left(763.2 \text{ K} \right)}{\pi \left(64 \text{ amu} \right) \left(1.66 \times 10^{-24} \text{ g/amu} \right)} \right]^{1/2} = 5.0 \times 10^4 \text{ cm/s}$$

$$D_b = 1/3 \left(1.5 \times 10^{-5} \text{ cm} \right) \left(5.0 \times 10^4 \text{ cm/s} \right) = 0.25 \text{ cm}^2/s$$

So, at high SV: $\delta = D \Big/ k_g = \dfrac{0.25 \text{ cm}^2/s}{\left(4.5 \times 10^4 \text{ cm/h} \right) \left(1 \text{ h/3600 s} \right)} = 0.020 \text{ cm}$

and at low SV: $\delta = \dfrac{0.25 \text{ cm}^2/s}{\left(2.8 \times 10^4 \text{ cm/h} \right) \left(1 \text{ h/3600 s} \right)} = 0.032 \text{ cm}$

at high SV: $\Delta C/\delta = \dfrac{2.6 \times 10^{-7} \text{ mole/cm}^3}{0.020 \text{ cm}} = 1.3 \times 10^{-5} \text{ mole/cm}^4$

at low SV: $\Delta C/\delta = \dfrac{2.9 \times 10^{-7} \text{ mole/cm}^3}{0.032 \text{ cm}} = 9.0 \times 10^{-6} \text{ mole/cm}^4$

c)   Nonisothermal situation, assume $C_s \cong C_o$

From Eq. 4.51 $\overline{\eta} = \dfrac{k_s}{k_o} \left(\dfrac{C_s}{C_o} \right)^n = \dfrac{k_s}{k_o}$

and $\dfrac{k_s}{k_o} = e^{-E/RT \, (1/t - 1)} \qquad \text{(Eq. 4.55)}$

also, $t = T_s / T_o = 1 + \overline{\beta} \cdot \overline{\eta} Da_o \qquad \text{(Eq. 4.56)}$

where $\overline{\beta} = \dfrac{-\Delta H_R^o C_o k_g}{T_o \, h}$

$\Delta H_R^o = -23.6 \text{ kcal/mole SO}_2 \qquad \text{(CRC Handbook)}$

$$\bar{\beta} = -\frac{\left(\dfrac{-23600 \text{ cal}}{\text{mole SO}_2}\right)\left(\dfrac{9.58 \times 10^{-7} \text{ mole SO}_2}{\text{cm}^3}\right)\left(\dfrac{9 \times 10^3 \text{ cm}^3 \text{ K}}{\text{cal}}\right)}{763.2 \text{ K}} = 0.27$$

Then $\Delta T = T_s - T_o = \bar{\beta} \cdot \bar{\eta} Da_o T_o$, and the $\bar{\beta} \cdot \bar{\eta} Da_o$ values are listed in Table 1.

d)   The non-isothermal effectiveness factor is

$$\bar{\eta}' = \frac{k_s}{k_o} = \frac{\Re_s}{\Re_o} = e^{-E/RT_o(1/t-1)} = \left(1 - \bar{\eta} Da_o\right) e^{-E/RT_o} \left(-\bar{\beta} \cdot \bar{\eta} Da_o\right)\left(1 + \bar{\beta} \cdot \bar{\eta} Da_o\right) \quad \text{(Eq. 4.57)}$$

where $E/RT_o = \dfrac{-30000 \text{ cal/mole}}{(1.987 \text{ cal/mole} \cdot \text{K})(763.2 \text{ K})} = -19.78$

The $\bar{\eta}'$ values are also given in Table 1.

Table 1

$\Re$ (mol/h·cm³)	k_g (cm/h)	$\bar{\eta} Da_o$	$\bar{\eta}$ (isothermal)	ΔC (mole/cm³)	$\bar{\beta} \cdot \bar{\eta} Da_o$	ΔT (K)	$\bar{\eta}'$ (noniso-thermal)
0.215	4.5x10⁴	0.27	0.73	2.6x10⁻⁷	0.072	55	3.8
0.204	3.8x10⁴	0.30	0.70	2.9x10⁻⁷	0.080	61	4.3
0.194	3.4x10⁴	0.32	0.68	3.0x10⁻⁷	0.085	65	4.7
0.153	2.8x10⁴	0.30	0.70	2.9x10⁻⁷	0.080	61	4.3

Problem 4.2 Solution

Using the Thiele modulus:

$$\eta = \frac{\mathfrak{R}}{\mathfrak{R}_o} = k_{meas}/k_o = 3/\phi\left(\frac{1}{\tanh \phi} - \frac{1}{\phi}\right) \quad \text{and} \quad \phi = R_p k_o^{1/2}/D_{eff}^{1/2} \Rightarrow$$

$$k_o = \frac{\phi}{3}\frac{k_{meas}}{\left(1/\tanh \phi - \frac{1}{\phi}\right)} = \phi^2 D_{eff}/R_p^2 \Rightarrow \phi\left(1/\tanh \phi - \frac{1}{\phi}\right) = \frac{k_{meas}\,\mathfrak{R}^2}{3\,D_{eff}}$$

$$R_p = \bar{r}_p = \frac{2\,V_g}{S_g} = 2\left(\frac{0.42 \text{ cm}^3/\text{g}}{(420 \text{ m}^2/\text{g})(100 \text{ cm/m})^2}\right) = 2.0\text{x}10^{-7} \text{ cm}$$

$$D_{eff} \cong D_{Kn} = 1/3\,d_p\,\bar{v} = 1/3\left(4.0\text{x}10^{-7} \text{ cm}\right)\left[\frac{8(1.38\text{x}10^{-16} \text{ erg/K})(685 \text{ K})}{\pi(120 \text{ amu})(1.66\text{x}10^{-24} \text{ g/amu})}\right]^{1/2} = 4.63\text{x}10^{-3} \frac{\text{cm}^2}{\text{s}}$$

$$\bar{v} = \left[\frac{8k_B T}{\pi m}\right]^{1/2} = \left[\frac{8RT}{\pi(MW)}\right]^{1/2} = 3.48\text{x}10^4 \text{ cm/s} \qquad (\text{Eq. 4.76})$$

$$\phi\left(\frac{1}{\tanh \phi} - \frac{1}{\phi}\right) = \frac{(1.49 \text{ cm}^3/\text{s}\cdot\text{g})(1.14 \text{ g/cm}^3)(0.35 \text{ cm}/2)^2}{3(4.6\text{x}10^{-3} \text{ cm}^2/\text{s})} = 3.75 \cong 3.8 \quad \text{so}$$

$$\frac{\phi}{\tanh \phi} = 4.8 \Rightarrow \phi = 4.8 \quad, \text{ and } \quad \eta = \frac{3}{4.8}\left(1 - \frac{1}{4.8}\right) = 0.495 \cong 0.5$$

Problem 4.3 Solution

$$2\,SO_2 + O_2 \Rightarrow 2\,SO_3$$

Use Weisz-Prater criterion :
$$N_{W\text{-}P} = \frac{\Re\,R_p^2}{C_s\,D_{eff}} \le 0.3$$

Assume $C_s = C_o = \dfrac{n}{V} = \dfrac{P}{RT} = \dfrac{2/3\,(2\ atm)}{\left(\dfrac{82.06\ atm\cdot cm^3}{mole\cdot K}\right)(673.2\ K)} = 2.4x10^{-5}\ \dfrac{mole}{cm^3}$

$$D_{eff} = \frac{1}{3}\bar{v}d_p = \frac{1}{3}\left(3x10^4\ cm/s\right)\left(1.2x10^{-6}\ cm\right) = 1.2x10^{-2}\ cm^2/s$$

$$R_p \le \left[\frac{0.3\,C_s\,D_{eff}}{\Re}\right]^{1/2} = \left[\frac{0.3\left(2.4x^{-5}\ mol/cm^3\right)\left(1.2x10^{-2}\ cm^2/s\right)}{\left(2\ mole\ SO_2/s\cdot\ell\right)\left(1\ \ell/1000\ cm^3\right)}\right]^{1/2} = 6.6x10^{-3}\ cm \Rightarrow$$

maximum particle diameter $= 1.3x10^{-2}\ cm$

Problem 4.4 Solution

Weisz-Prater criterion: $N_{W-P} = \dfrac{\Re R_p^2}{C_s D_{eff}} < 0.3 \Rightarrow$ no significant pore diffusion effect

$\Re = (0.0956 \text{ g mol } SO_2/h \cdot g \text{ cat})(1.6 \text{ g/cm}^3)(h/3600 \text{ s}) = 4.25 \times 10^{-5} \text{ mole/s} \cdot \text{cm}^3$

$R_p = 1.6 \text{ mm} = 0.16 \text{ cm}$

Assume $C_s = C_o = 9.58 \times 10^{-7} \text{ mole/cm}^3$

$$D_{eff} \cong D_{Kn} = \frac{1}{3}\bar{v}d_p = \frac{1}{3}\left[\frac{8(1.38 \times 10^{-16} \text{ erg/K})(763 \text{ K})}{\pi(64 \text{ amu})(1.66 \times 10^{-24} \text{ g/amu})}\right]^{1/2}[100 \times 10^{-8} \text{ cm}] = 1.67 \times 10^{-2} \text{ cm}^2/\text{s}$$

$$N_{W-P} = \frac{(4.25 \times 10^{-5} \text{ mole/s} \cdot \text{cm}^3)(0.16 \text{ cm})^2}{(9.58 \times 10^{-7} \text{ mole/cm}^3)(1.67 \times 10^{-2} \text{ cm}^2/\text{s})} = 68 >> 6$$

Mass transport limitations due to pore diffusion are definitely present.

Problem 4.5 Solution

$$N_{W-P} = \Re\, R_p^2 / C_s D_{eff} \cong \Re\, R_p^2 / C D_{Kn}$$

$$\Re = \left(\frac{1.99\ \mu\text{mole Bz}}{\text{s g cat}}\right)\left(\frac{0.60\ \text{g cat}}{\text{cm}^3}\right) = \frac{1.2\ \mu\ \text{mole Bz}}{\text{s cm}^3}$$

$$C_s = C_o = P/RT = \frac{(50\ \text{Torr Bz})(1\ \text{atm}/760\ \text{Torr})}{\left(\dfrac{82.06\ \text{atm}\cdot\text{cm}^3}{\text{g mole K}}\right)(413\ \text{K})} = 1.9\text{x}10^{-6}\ \frac{\text{mole}}{\text{cm}^3}$$

$D_{eff} \cong D_{Kn}$ because $\lambda \gg d_p$ (average pore diameter)

$D_{Kn} = 1/3\ \bar{v} d_p$ and let $d_p = 250\ \overset{\circ}{\text{A}} = 25\ \text{nm}$

$$\bar{v} = (8\,k_B T/\pi m)^{1/2} = \left[\frac{8\left(1.38\text{x}10^{-16}\ \text{erg/K}\right)(413\ \text{K})}{\pi\left(78\ \text{amu}\right)\left(1.66\text{x}10^{-24}\ \text{g/amu}\right)}\right]^{1/2} = 3.35\text{x}10^4\ \frac{\text{cm}}{\text{s}}\quad\text{(Eq. 4.76)}$$

$$D_{Kn} = 1/3\left(\frac{3.35\text{x}10^4\ \text{cm}}{\text{s}}\right)(25\ \text{nm})\left(\frac{1\ \text{cm}}{10^7\ \text{nm}}\right) = 2.8\text{x}10^{-2}\ \frac{\text{cm}^2}{\text{s}}\quad\text{(Eq. 4.78)}$$

(a) With Pd in mesopores = 25 nm:

Worst case: $R_p = 500\ \mu$; $N_{W-P} = \dfrac{\left(1.2\text{x}10^{-6}\ \text{mole Bz/s cm}^3\right)\left(500\text{x}10^{-4}\ \text{cm}\right)^2}{\left(1.9\text{x}10^{-6}\ \text{Bz/cm}^3\right)\left(2.8\text{x}10^{-2}\ \text{cm}^2/\text{s}\right)} = 0.056$

Best case:

$$R_p = 10\ \mu\ ;\ N_{W-P} = \frac{\left(1.2\text{x}10^{-6}\ \text{mole Bz s}\cdot\text{cm}^3\right)\left(10\text{x}10^{-4}\ \text{cm}\right)^2}{\left(1.9\text{x}10^{-6}\ \text{mole Bz/cm}^3\right)\left(2.8\text{x}10^{-2}\ \text{cm}^2/\text{s}\right)} = 2.3\text{x}10^{-5}$$

(b) With Pd in micropores = 0.9 nm:

$$D_{Kn} = 1/3\left(3.35\text{x}10^4\ \text{cm/s}\right)(0.9\ \text{nm})\left(1\ \text{cm}/10^7\ \text{nm}\right) = 1.0\text{x}10^{-3}\ \text{cm}^2/\text{s}$$

Worst case: $R_p = 500\ \mu$; $N_{W-P} = \dfrac{\left(1.2\text{x}10^{-6}\ \text{mole Bz/s}\cdot\text{cm}^3\right)\left(500\text{x}10^{-4}\ \text{cm}\right)^2}{\left(1.9\text{x}10^{-6}\ \text{mole Bz/cm}^3\right)\left(1.0\text{x}10^{-3}\ \text{cm}^2/\text{s}\right)} = 1.6$

Best case: $R_p = 10\ \mu$; $N_{W-P} = \dfrac{\left(1.2\text{x}10^{-6}\ \text{mole Bz/s}\cdot\text{cm}^3\right)\left(10\text{x}10^{-4}\ \text{cm}\right)^2}{\left(1.9\text{x}10^{-6}\ \text{mole Bz/cm}^3\right)\left(1.0\text{x}10^{-3}\ \text{cm}^2/\text{s}\right)} = 6.3\text{x}10^{-4}$

Problem 4.6 Solution

$$N_{W-P} = \Re R_p^2 / C_s D_{eff} \cong r_{CO_2} \rho_{cat} R_p^2 / (C_s D_{Kn}) \text{ because } D_{Kn} < D_{eff}$$

(1) $C_{CO_2} = \dfrac{n}{V} = \dfrac{P}{RT} = \dfrac{(200 \text{ Torr})\left(\dfrac{1 \text{ atm}}{760 \text{ Torr}}\right)}{\left(82.06 \dfrac{\text{atm} \cdot \text{cm}^3}{\text{g mole K}}\right)(723 \text{ K})} = 4.4 \times 10^{-6} \dfrac{\text{mole } CO_2}{\text{cm}^3}$

(2) $\bar{v} = \left(\dfrac{8 k_B T}{\pi m}\right)^{1/2} = \left[\dfrac{8(1.38 \times 10^{-16} \text{ erg/K})(723 \text{ K})}{\pi(44 \text{ amu})(1.66 \times 10^{-24} \text{ g/amu})}\right]^{1/2} = 5.9 \times 10^4 \text{ cm/s}$ (Eq. 4.76)

a)

$$D_{Kn} = 1/3\,\bar{v}\,d_p = 1/3\,(5.9 \times 10^4 \text{ cm/s})(18 \text{ nm})(1 \text{ cm}/10^7 \text{ nm}) = 3.5 \times 10^{-2} \text{ cm}^2/\text{s} \quad (\text{Eq. 4.78})$$

$$N_{W-P} = \dfrac{(42.6 \times 10^{-6} \text{ mole/s} \cdot \text{g})(1 \text{ g/cm}^3)(0.01 \text{ cm})^2}{(4.4 \times 10^{-6} \text{ mole/cm}^3)(3.5 \times 10^{-2} \text{ cm}^2/\text{s})} = \underline{0.028} \quad (\text{No problem})$$

b) $D_{Kn} = 1/3(5.9 \times 10^4 \text{ cm/s})(14 \text{ nm})(1 \text{ cm}/10^7 \text{ nm}) = 2.8 \times 10^{-2} \text{ cm}^2/\text{s}$

$$N_{W-P} = \dfrac{(9.6 \times 10^{-6} \text{ mole/s} \cdot \text{g})(1 \text{ g/cm}^3)(0.01 \text{ cm})^2}{(4.4 \times 10^{-6} \text{ mole/cm}^3)(2.8 \times 10^{-2} \text{ cm}^2/\text{s})} = \underline{7.8 \times 10^{-3}} \quad (\text{No problem})$$

c) $D_{Kn} = 1/3(5.9 \times 10^4 \text{ cm/s})(20 \text{ nm})(1 \text{ cm}/10^7 \text{ nm}) = 3.9 \times 10^{-2} \text{ cm}^2/\text{s}$

$$N_{W-P} = \dfrac{(118 \times 10^{-6} \text{ mole/s} \cdot \text{g})(1 \text{ g/cm}^3)(0.01 \text{ cm})^2}{(4.4 \times 10^{-6} \text{ mole/cm}^3)(3.9 \times 10^{-2} \text{ cm}^2/\text{s})} = \underline{6.9 \times 10^{-3}} \quad (\text{No problem})$$

d) $D_{Kn} = 1/3(5.9 \times 10^4 \text{ cm/s})(20 \text{ nm})(1 \text{ cm}/10^7 \text{ nm}) = 3.9 \times 10^{-2} \text{ cm}^2/\text{s}$

$$N_{W-P} = \dfrac{(237 \times 10^{-6} \text{ mole/s} \cdot \text{g})(1 \text{ g/cm}^3)(0.01 \text{ cm})^2}{(4.4 \times 10^{-6} \text{ mole/cm}^{-3})(3.9 \times 10^{-2} \text{ cm}^2/\text{s})} = \underline{0.14} \quad (\text{Possible concern})$$

Problem 4.7 Solution

Thiele modulus
$$\phi = \frac{R_p k^{1/2} C_s^{\frac{n-1}{2}}}{D_{\text{eff}}^{1/2}}$$

Rate $= k' P_{H_2} = k C_{H_2} = \left(\dfrac{1.99\ \mu\text{mole Bz}}{\text{s} \cdot \text{g}}\right)\left(0.60\ \text{g}/\text{cm}^3\right) = 1.19\text{x}10^{-6}\ \dfrac{\text{mole Bz}}{\text{s} \cdot \text{cm}^3}$

$C_{H_2} = \dfrac{n}{V} = \dfrac{P}{RT} = \dfrac{\left(710/760\ \text{atm}\right)}{\left(\dfrac{82.06\ \text{cm}^3 \cdot \text{atm}}{\text{g mole} \cdot \text{K}}\right)\left(413\text{K}\right)} = 2.76\text{x}10^{-5}\ \dfrac{\text{mole}}{\text{cm}^3}$

$k = \dfrac{\text{Rate}}{C_{H_2}} = \dfrac{1.19\text{x}10^{-6}\ \text{mole}/\text{s} \cdot \text{cm}^3}{2.76\text{x}10^{-5}\ \text{mole}/\text{cm}^3} = 0.0432\ \text{s}^{-1}$

$D_{H_2} = 1/3\, \overline{v}_{H_2} \lambda$

$\overline{v}_{H_2} = \left[\dfrac{8\left(1.38\text{x}10^{-16}\ \text{erg/K}\right)\left(413\text{K}\right)}{\pi\left(2\ \text{amu}\right)\left(1.66\text{x}10^{-24}\ \text{g/amu}\right)}\right]^{1/2} = 2.1\text{x}10^5\ \text{cm/s}$

$\lambda = \dfrac{\left(\dfrac{82.06\ \text{cm}^3 \cdot \text{atm}}{\text{gmole} \cdot \text{K}}\right)\left(413\text{K}\right)\left(\dfrac{1\ \text{mole}}{6.02\text{x}10^{23}\ \text{molecules}}\right)}{\sqrt{2}\ \pi\left(2.4\text{x}10^{-8}\ \text{cm}\right)^2\left(1\ \text{atm total P}\right)} = 2.2\ \text{x}\ 10^{-5}\ \text{cm}$

$D_{H_2} = 1/3\left(2.1\text{x}10^5\ \text{cm/s}\right)\left(2.2\text{x}10^{-5}\ \text{cm}\right) = 1.5\ \text{cm}^2/\text{s}$

$D_{\text{eff}} \cong D_{Kn} = 1/3\, \overline{v} d_p \quad$ where $\quad d_p = $ pore diameter

Case 1: Smallest particle, all Pd in mesopores:

$$D_{Kn} = 1/3 \left(2.1 \times 10^5 \text{ cm/s}\right)\left(25 \times 10^{-9} \text{ m}\right)\left(10^2 \text{ cm/m}\right) = 0.175 \text{ cm}^2/\text{s}$$

$$\phi = \frac{\left(10 \times 10^{-6} \text{ m}\right)\left(10^2 \text{ cm/m}\right)\left(0.0432/\text{s}\right)^{1/2}}{\left(0.175 \text{ cm/s}\right)^{1/2}} = 5.0 \times 10^{-4}$$

Case 2: Largest particle, all Pd in mesopores:

$$\phi = \frac{\left(500 \times 10^{-6} \text{ m}\right)\left(10^2 \text{ cm/m}\right)\left(0.0432/\text{s}\right)^{1/2}}{\left(0.175 \text{ cm/s}\right)^{1/2}} = 0.025$$

Case 3: Largest particle, all Pd in micropores:

$$D_{Kn} = 1/3 \left(2.1 \times 10^5 \text{ cm/s}\right)\left(0.9 \times 10^{-9} \text{ m}\right)\left(10^2 \text{ cm/m}\right) = 6.3 \times 10^{-3} \text{ cm}^2/\text{s}$$

$$\phi = \frac{\left(500 \times 10^{-6} \text{ m}\right)\left(10^2 \text{ cm/m}\right)\left(0.0432/\text{s}\right)^{1/2}}{\left(6.3 \times 10^{-3} \text{ cm}^2/\text{s}\right)^{1/2}} = 0.13$$

No concern – in all cases η is essentially unity.

Problem 4.8 Solution

Use Weisz-Prater criterion.

$$N_{w-p} = \frac{\Re R_p^2}{C_s\, D_{eff}} \quad << \quad 0.6 \text{ (or 0.3)} \qquad R_p = \frac{1}{2}\left(14.9 \times 10^{-3}\ cm\right) \ , \quad \lambda = d_p$$

$$D_{eff} = \frac{1}{3}\bar{v}d_p = \frac{1}{3}\left[\frac{8\left(1.38 \times 10^{-16}\ erg/K\right)\left(493\ K\right)}{\pi\left(30\ amu\right)\left(1.66 \times 10^{-24}\ g/amu\right)}\right]\left(60 \times 10^{-8}\ cm\right) = 1.18 \times 10^{-2}\ cm^2/s \quad \text{(Eq. 4.78)}$$

Assume 1 atm:

$$C_s = C = \frac{n}{V} = \frac{P}{RT} = \frac{\left(9\ Torr\right)\left(1\ atm/760\ Torr\right)}{\left(82.06\ \dfrac{cm^3 \cdot atm}{gmol \cdot K}\right)\left(493\ K\right)} = 2.93 \times 10^{-7}\ \frac{mole\ CH_2O}{cm^3}$$

$$N_{w-p} = \frac{\left[1.4 \times 10^{-7}\ mole/cm^3 \cdot s\right]\left[7.45 \times 10^{-3}\ cm\right]^2}{\left[2.93 \times 10^{-7}\ mole/cm^3\right]\left[1.18 \times 10^{-2}\ cm^2/s\right]} = 2.25 \times 10^{-3} \quad << 0.3$$

Yes, it's free of diffusion effects.

Problem 5.1 Solution

$$K = \frac{[O-*]^2 [S-*]}{P_{SO_2} [*]^3} \quad , \quad L = [O-*] + [S-*] + [*] \quad \text{and} \quad [O-*] = 2[S-*]$$

$$[*] = L - [O-*] - [S-*] = L - 3[S-*]$$

$$K = \frac{(2[S-*])^2 [S-*]}{P_{SO_2} [*]^3} = \frac{4[S-*]^3}{P_{SO_2} (L-3[S-*])^3}$$

$$K (L-3[S-*])^3 = 4[S-*]^3 / P_{SO_2} \Rightarrow K^{1/3} (L-3[S-*]) = \sqrt[3]{4} [S-*] / P_{SO_2}^{1/3}$$

$$LK^{1/3} P_{SO_2}^{1/3} - 3[S-*] K^{1/3} P_{SO_2}^{1/3} = \sqrt[3]{4} [S-*]$$

$$LK^{1/3} P_{SO_2}^{1/3} = \left[3 K^{1/3} P_{SO_2}^{1/3} + \sqrt[3]{4} \right] [S-*]$$

$$[S-*] = \frac{LK^{1/3} P_{SO_2}^{1/3}}{\sqrt[3]{4} + 3 K^{1/3} P_{SO_2}^{1/3}} = \frac{L (K/4)^{1/3} P_{SO_2}^{1/3}}{1 + 3 (K/4)^{1/3} P_{SO_2}^{1/3}} = \frac{L \left(K^{1/3} P_{SO_2}^{1/3} \right)}{1 + 3 K^{1/3} P_{SO_2}^{1/3}}$$

$$\theta_s = \frac{[S-*]}{L} = \frac{K^{1/3} P_{SO_2}^{1/3}}{1 + 3 K^{1/3} P_{SO_2}^{1/3}} \quad , \quad \theta_{s\,max} = 1/3$$

$$\theta = 3 \theta_s = \frac{3 K^{1/3} P_{SO_3}^{1/3}}{1 + 3 K^{1/3} P_{SO_2}^{1/3}}$$

Problem 5.2 Solution

To test single-site adsorption, plot $\dfrac{P}{n} = \dfrac{1}{Kn_m} + \dfrac{P}{n_m}$ (Eq. 5.27), whereas to test

dual (double)-site adsorption, plot $\dfrac{P^{1/2}}{n} = \dfrac{1}{K^{1/2}n_m} + \dfrac{P^{1/2}}{n_m}$ (Eq. 5.28). The single-site

equation gives a reasonable linear fit, whereas the dual-site equation is curved with

negative slopes (See Figures 1 and 2, respectively).

From the slopes and intercepts of the plots in Figure 1:

T(K)	K (atm^{-1})	n_m (µmole g^{-1})	or	$V_m \left(cm^3_{STP}\ g^{-1}\right)^a$
343	181	1380		30.9
363	87	1180		26.6
383	59	772		17.1
403	28	719		16.1

Use Eq. 5.38 and plot ℓn K vs. $1/T$ (Figure 3)

Slope $= 4.13 \times 10^3 = -\Delta H^{\circ}_{ad}/R$, so $\Delta H^{\circ}_{ad} = -8.2\ \dfrac{kcal}{mole}$ or $-34\ \dfrac{kJ}{mole}$

Intercept $= -6.85 = \Delta S^{\circ}_{ad}/R$, so $\Delta S^{\circ}_{ad} = -14\ \dfrac{cal}{mole \cdot K}$ or $-59\ \dfrac{J}{mole \cdot K}$

(a)

$V_m = nRT/P = \left(n\ g\ mole\right)\dfrac{\left(82.06\ cm^3 \cdot atm/g\ mole \cdot K\right)\left(273.2\ K\right)}{1\ atm} = n\ \left(2.242 \times 10^4\ cm^3\right)$

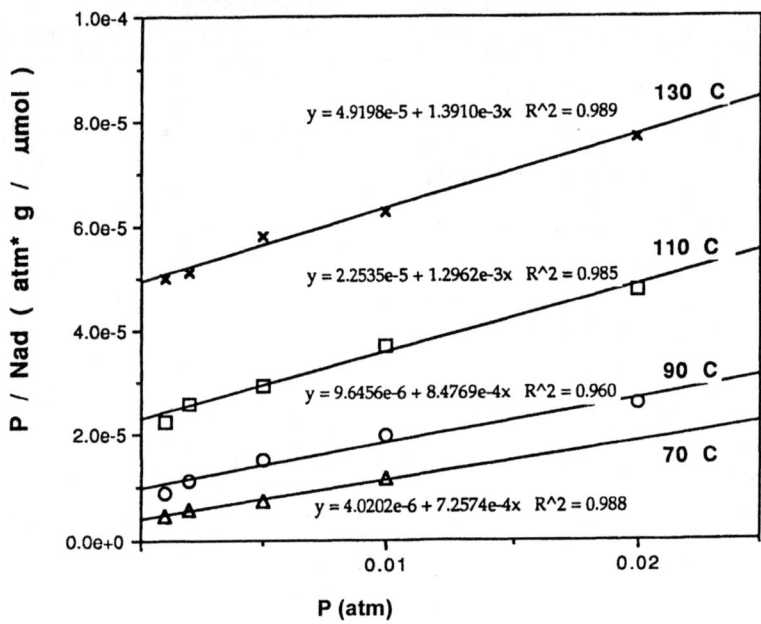

Benzene Adsorption On Silica Gel
Single Site Langmuir Isotherm

130 C
$y = 4.9198e-5 + 1.3910e-3x \quad R^2 = 0.989$

110 C
$y = 2.2535e-5 + 1.2962e-3x \quad R^2 = 0.985$

90 C
$y = 9.6456e-6 + 8.4769e-4x \quad R^2 = 0.960$

70 C
$y = 4.0202e-6 + 7.2574e-4x \quad R^2 = 0.988$

P / Nad (atm* g / μmol)

P (atm)

Figure 1

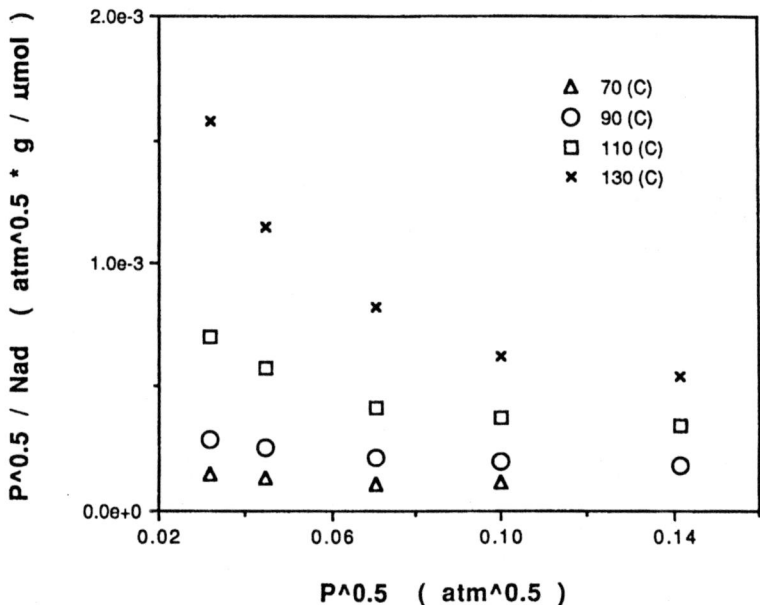

Figure 2

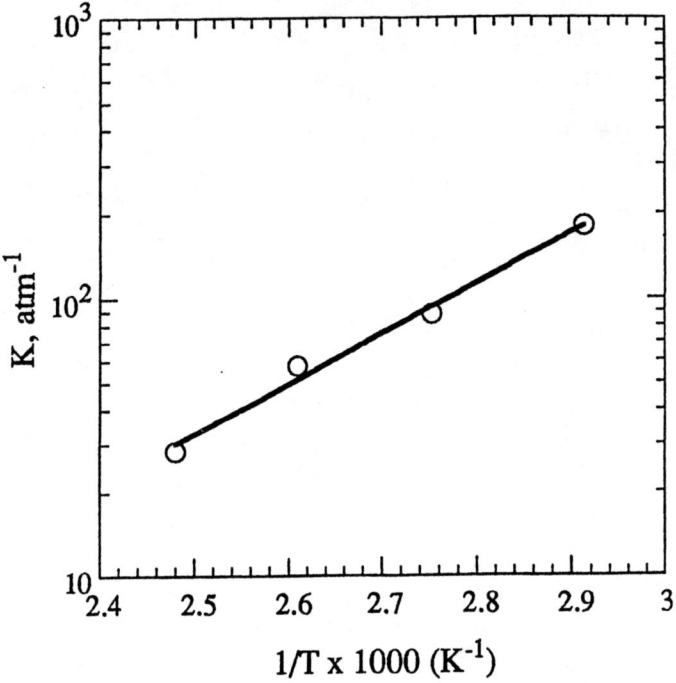

Plot of K vs. 1/T for Pure Benzene Adsorption

Figure 3

Problem 5.3 Solution

To test isotherms:

For single-site Langmuir isotherm, plot P/n vs. P (Eq. 5.71) – See Figure 4

For Freundlich isotherm, plot $\ell n\ n$ vs. $\ell n\ P$ (Eq. 5.73) – See Figure 5

For Temkin isotherm, plot n vs. $\ell n\ P$ (Eq. 5.74) – See Figure 6

The Freundlich isotherm gives the most linear best fit.

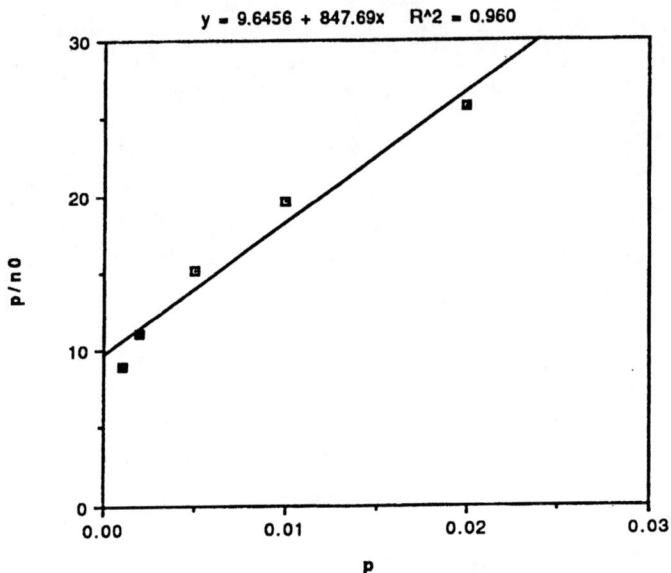

LANGMUIR SINGLE SITE ADSORPTION ISOTHERM

y = 9.6456 + 847.69x   R^2 = 0.960

Figure 4

FREUNDLICH ADSORPTION ISOTHERM

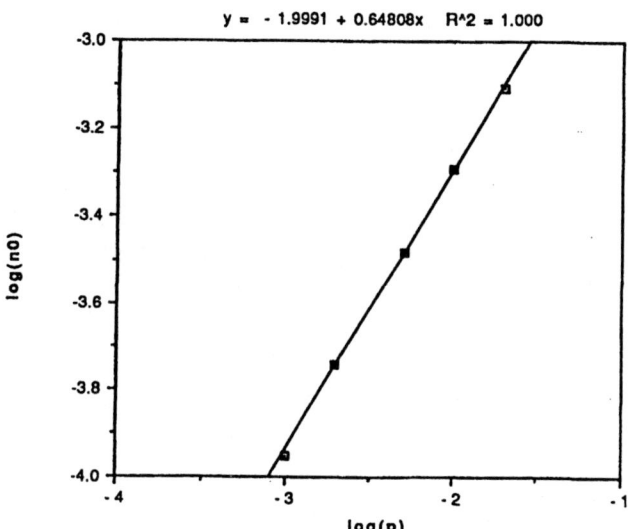

y = - 1.9991 + 0.64808x   R^2 = 1.000

Figure 5

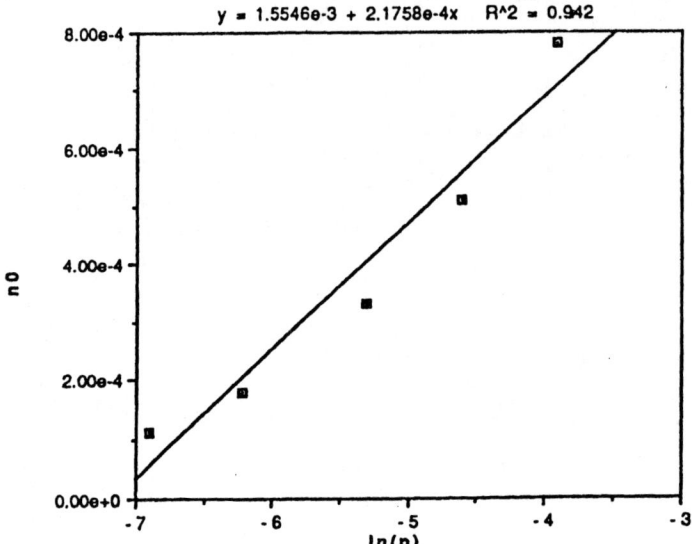

Figure 6

Problem 5.4 Solution

$$A + 2S \overset{K}{\rightleftharpoons} S\text{-}A\text{-}S$$

$L = S + 2 \,(SAS)$ $\qquad\qquad K = [SAS] / [A][S]^2$

$S = L - 2\,[SAS] \quad \Rightarrow \quad S^2 = L^2 - 4L\,[SAS] + 4\,[SAS]^2$

$[SAS] = [A]\,[S]^2\,K = K\,[A]\,(L^2 - 4L\,[SAS] + 4\,[SAS]^2)$

$$\underbrace{4\,K\,[A]\,[SAS]^2}_{a} - \underbrace{(4LK[A]+1)}_{b}\,[SAS] + \underbrace{L^2K\,[A]}_{c} = 0 \qquad x = \frac{-b \pm \sqrt{b^2 - 4ac}}{2a} \quad \text{for } ax^2 + bx + c = 0$$

$$[SAS] = \frac{(1 + 4\,LK\,[A]) \pm \sqrt{1 + 8\,LK\,[A] + 16\,L^2\,K^2\,[A]^2 - 16\,L^2\,K^2\,[A]^2}}{8\,K\,[A]}$$

$$[SAS] = \frac{(1 + 4\,LK\,[A]) - \sqrt{1 + 8LK\,[A]}}{8K\,[A]} \qquad\qquad \begin{array}{l} [SAS] = 0 \quad @ \;\; [A] = 0 \\[4pt] [SAS] \to L/2 \;\; \text{as} \;\; [A] \to \infty \end{array}$$

OR, alternatively

$$2[SAS]/L = \theta_A \quad S/L = 1 - 2[SAS]/L \;\; \Rightarrow \theta = 1 - \theta_A \;\; \& \;\; K = \frac{L\theta_A}{L^2[A]\theta^2}$$

$$LK\,[A]\theta^2 = \theta_A \;\; \Rightarrow \;\; (1 - \theta_A)^2\,LK\,[A] = \theta_A$$

$$LK\,[A]\theta_A^2 - (2\,LK\,[A] + 1)\,\theta_A + LK\,[A] = 0$$

$$\theta_A = \frac{2\,LK\,[A] + 1 \;\; \pm \;\; \sqrt{4L^2K^2\,[A]^2 + 4LK\,[A] + 1 - 4L^2K^2[A^2]}}{2\,LK\,[A]} \;\; \Rightarrow$$

$$\theta_A = \frac{1 + 2\,LK\,[A] \pm \sqrt{1 + 4LK\,[A]}}{2LK\,[A]} \;\; \Rightarrow \;\; \theta_A = \frac{1 + 2\,LK\,[A] - \sqrt{1 + 4LK\,[A]}}{2LK\,[A]}$$

$\theta_A = 0 \;\; @ \;\; [A] = 0 \quad \text{and} \quad \theta_A \to 1 \quad \text{as} \quad [A] \to \infty$

Problem 6.1 Solution

$k = Ae^{-E/RT}$ so $\ln k = \ln A - E/RT$

Plot $\ln k$ vs. $1/T$

T (K)	k
823	1.3×10^{-5}
833	2.3×10^{-5}
843	4.1×10^{-5}
853	6.2×10^{-5}
863	11.5×10^{-5}
873	17.7×10^{-5}
883	28.6×10^{-5}
893	46.2×10^{-5}
903	70.8×10^{-5}

$$\text{Slope} \; = \; -E/R \; = \; -37,200 \; \Rightarrow \; E \; = \; 73.9 \, \frac{\text{kcal}}{\text{mole}} \; = \; 309 \, \frac{\text{kJ}}{\text{mole}}$$

$$\text{Intercept} \; = \; \ln A \; = \; 33.86 \; \Rightarrow \; A \; = \; 5.1 \times 10^{14} \; s^{-1}$$

Problem 6.2 Solution

a) $\quad r = v \left[X^{\ddagger} \right] = k_B T/h \, \dfrac{Q_{N-N}}{Q_N^2} \, e^{-E/RT} \left[N \cdot \right]^2$

pre-exponential factor $= k_o = \left(k_B T/h \right) \dfrac{f_{tr}^3 f_{rot}^2}{f_{tr}^6}$ because a N atom has no rotational or vibrational

modes. Therefore, $k_o = \left(k_B T/h \right) \dfrac{(aT)^{3/2} (bT)}{(cT)^3} \Rightarrow k_o \; \alpha \; T^{1/2}$

b) $\quad Q_{CO} = f_{tr}^3 \, f_{rot}^2 \, f_{vib} \;$ and $\; Q_{H_2O} = f_{tr}^3 \, f_{rot}^3 \, f_{vib}^3$

Q_{COOH_2} has $3 \times 5 = 15$ modes, with $15 - 7 = 8$ degrees of vibrational freedom; therefore

$$k_o = \left(k_B T/h \right) \dfrac{f_{tr}^3 \, f_{rot}^3 \, f_{vib}^8}{f_{tr}^6 \, f_{rot}^5 \, f_{vib}^4} \cong \left(k_B T/h \right) \dfrac{1}{f_{tr}^3 \, f_{rot}^2} \quad \text{because } f_{vib} \cong 1$$

so $\; k_o \; \alpha \left(k_B T/h \right) \dfrac{1}{(aT)^{3/2} (bT)} \;$ or $\; k_o \; \alpha \; T^{-3/2}$

c) $\quad Q_{H_2} \cong Q_{Cl_2} = f_{tr}^3 \, f_{rot}^2 \, f_{vib}$; assume non-linear activated complex, so degrees of vibrational freedom $= 3N - 7 = 5$, and $Q^{\ddagger} = Q_{H_2Cl_2} = f_{tr}^3 \, f_{rot}^3 \, f_{vib}^5$. Therefore,

$$k_o = \left(k_B T/h \right) \dfrac{f_{tr}^3 \, f_{rot}^3 \, f_{vib}^5}{f_{tr}^6 \, f_{rot}^4 \, f_{vib}^2} = \left(k_B T/h \right) \dfrac{1}{f_{tr}^3 \, f_{rot}} \quad \text{because } f_{vib} \cong 1, \text{ so } k_o \; \alpha \left(k_B T/h \right) \dfrac{1}{(aT)^{3/2} (bT)^{1/2}}$$

and $\; k_o \; \alpha \; T^{-1}$

Problem 6.3 Solution

$$H_2 + Br_2 \implies 2\,HBr$$

1. $$r = \frac{d\,[HBr]}{dt} = k_1[Br\cdot][H_2] + k_2[H\cdot][Br_2] - k_{-1}[H\cdot][HBr]$$

(1) Use SSA on H· radical: $\dfrac{d\,[H\cdot]}{dt} = 0 = k_1\,[Br\cdot][H_2] - k_2\,[H\cdot][Br_2] - k_{-1}\,[H\cdot][HBr]$

(2) Use SSA on Br· radical:

$$\frac{d\,[Br\cdot]}{dt} = 0, = \text{ but an easier way is}: \; r_i = r_t \implies 2k_i\,[Br_2] = 2\,k_t\,[Br\cdot]^2 \implies$$

$$[Br\cdot] = (k_i/k_t)^{1/2}\,[Br_2]^{1/2}$$

From (1): $\quad [H\cdot] = \dfrac{k_1\,[Br\cdot][H_2]}{k_2\,[Br_2] + k_{-1}\,[HBr]}$, then

$$[H\cdot] = \frac{k_1(k_i/k_t)^{1/2}\,[Br_2]^{1/2}\,[H_2]}{k_2\,[Br_2] + k_{-1}\,[HBr]}$$

$$r = k_1(k_1/k_t)^{1/2}\,[Br_2]^{1/2}\,[H_2] + (k_2[Br_2] - k_{-1}[HBr])\left(\frac{k_1\,(k_i/k_t)^{1/2}\,[Br_2]^{1/2}\,[H_2]}{k_2\,[Br_2] + k_{-1}\,[HBr]}\right)$$

$$= \frac{(k_2\,[Br_2] + k_{-1}\,[HBr])(k_1(k_i/k_t)^{1/2}[Br_2]^{1/2}[H_2]) + (k_2\,[Br_2] - k_{-1}\,[HBr])(k_1\,(k_i/k_t)^{1/2}[Br_2]^{1/2}[H_2])}{k_2\,[Br_2] + k_{-1}\,[HBr]}$$

$$= \frac{2k_1 k_2 (k_i/k_t)^{1/2}[Br_2]^{3/2}[H_2]}{k_2\,[Br_2] + k_{-1}\,[HBr]} \qquad \text{or}$$

$$r = \frac{2k_1\,(k_i/k_t)^{1/2}[Br_2]^{1/2}[H_2]}{1 + \dfrac{k_{-1}}{k_2}\dfrac{[HBr]}{[Br_2]}}$$

Problem 6.4

$$CH_3CHO \Rightarrow CH_4 + CO$$

$$r = -\frac{d[CH_3CHO]}{dt} = r_2 = r_3 = \frac{d[CH_4]}{dt} = \frac{dCO}{dt}$$

at constant volume

$$r_i = r_t \Rightarrow r_1 = 2 r_4 \quad \text{so} \quad k_1 [CH_3CHO] = 2 k_4 [CH_3 \cdot]^2$$

$$r = k_2 [CH_3 \cdot][CH_3CHO] = k_2 (k_1/2 k_4)^{1/2} [CH_3CHO]^{3/2}, \text{ thus}$$

$$r = k_{app} [CH_3CHO]^{3/2} \quad \text{so} \quad E_{app} = E_2 + \frac{E_1}{2} - \frac{E_4}{2}$$

Problem 6.5 Solution

a) $Q_N > Q_O$ so   N end down, i.e., $\overset{\nearrow N=O}{Fe}$, $n = 3$.

For $\left(\eta^1 \mu_1\right)$: $Q_{oN} = \dfrac{Q_N}{(2 - 1/n)} = \dfrac{140}{(5/3)} = 84$ and $Q_{NO} = \dfrac{Q_{oN}^2}{\dfrac{Q_{oN}}{n'} + D_{NO}}$ (Table 6.5)

so $Q_{NO} = (0.6)^2 (140)^2 / \left[(0.6)(140) + 151\right] = 30$ kcal mole^{-1}

For $\left(\eta^1 \mu_2\right)$: $Q_{NO} = (0.6)^2 (140)^2 \Big/ \left[\dfrac{(0.6)(140)}{2} + 151\right] = 37$ kcal mole^{-1} $\Rightarrow$ $\overset{\displaystyle O}{\underset{\displaystyle Fe\!-\!Fe}{\overset{\|}{\underset{\diagdown}{\overset{N}{\diagup}}}}}$

(preferred)

b)   Orientation is $\eta^1 \mu_1$ with O end down, n = 4 .

$Q_{oO} = \dfrac{115}{(7/4)} = 65.7$ kcal mole^{-1} (Table 6.5)

$Q_{H_2O} = \dfrac{Q_{oO}^2}{Q_{oO} + D_{H_2O}} = \dfrac{(65.7)^2}{65.7 + 220} = 15$ kcal mole   (Table 6.7a)

c)   This is a symmetric molecule with A = CH, i.e., $\underset{Pt\!-\!\!-\!\!-\!\!-\!Pt}{\overset{H\diagdown\quad\diagup H}{\underset{}{C=C}}}$, n=3 .

For $\left(\eta^2 \mu_2\right)$: $D_{C\equiv C} = D_{HC\equiv CH} - 2\, D_{CH} = 392 - 2\,(81) = 230$ kcal mole^{-1} (Table 6.7b)

$Q_{oC} = Q_C / (2 - 1/3) = (3/5)(150) = 90$

$D_{H-C\equiv} = 230 + 81 = 311 = D_{C\equiv C} + D_{C-H}$

$Q_{HC\equiv CH} = \dfrac{(9/2)(90)^2}{3\,(90) + 8\,(311)} = 13$ kcal mole   (Eq. 6.46)

d) $Q_{CH_2} = \dfrac{Q_C^2}{Q_C + D_{CH_2}}$ for "strong" chemisorption (adsorbs in hollow site)

use $Q_C = 160 \text{ kcal mole}^{-1}$ on $Pd(111)$,   so

$Q_{CH_2} = \dfrac{(160)^2}{160 + 183} = 75 \text{ kcal mole}^{-1}$

Problem 6.6 Solution

a) $N_{ad} > O_{ad}$ regarding bond strength, so N end down. This is "weak" chemisorption, $n = 4$.

$$
\begin{array}{c}
O \\
\parallel \\
N \\
\diagdown \\
\text{Ni} \text{---} \text{Ni}
\end{array}
$$

For $\eta^1\mu_2 \Rightarrow \overset{\diagup}{\text{Ni}}\text{---}\text{Ni}$ so $n' = 2$,

$$Q_{NO} = \frac{Q_{oA}}{\dfrac{Q_{oA}}{n'} + D_{AB}} = \frac{Q_{oN}}{\dfrac{Q_{oN}}{2} + 151} \quad \text{(Table 6.6), so } Q_{NO} = \frac{(4/7)^2(135)^2}{\dfrac{(4/7)(135)}{2} + 151} = 31 \text{ kcal mole}^{-1}$$

b) This is "weak" chemisorption, $n = 3$,

$$
\begin{array}{c}
H\ H\ H\ H \\
\diagdown|\diagup\diagdown|\diagup \\
C \cdots\cdots C \\
\diagup \quad \diagdown \\
Pd \text{------} Pd
\end{array}
$$

$D_{C=C} = D_{CH_2CH_2} - 2\,D_{CH_2} = 538 - 366 = 172$, A equals $= CH_2$, and

(Table 6.7b)

$D_{=CH_2} = D_{CH_2} + D_{C=C} = 183 + 172 = 355$

$$Q_{A_2} = \frac{9/2\, Q_{oC}^2}{3\, Q_{oC} + 8\, D_{A_2}} = (9/2)(0.6)^2 \,(160)^2 / [3\,(0.6)(160) + 8\,(355)] = 13 \text{ kcal mole}^{-1}$$

c) This is "weak" chemisorption, $n = 3$, N end down, i.e.,

$$
\begin{array}{c}
H \\
| \\
H\text{-}N\text{-}H \\
| \\
Fe
\end{array}
$$

$Q_{NH_3} = \dfrac{Q_{oA}^2}{Q_{oA} + D_{AB}}$, from ref. 25, $D_{NH_3} = 279$ kcal mole^{-1},

so $Q_{NH_3} = \dfrac{(0.6)^2(140)^2}{(0.60)(140) + 279} = 19$ kcal mole^{-1}

d)   This is "strong" chemisorption

$$Q_{NH_2} = \frac{Q_N^2}{Q_N + D_{NH_2}}, \text{ from ref. 25, } D_{NH_2} = 169 \text{ kcal mole}^{-1},$$

$$\text{so } Q_{NH_2} = \frac{(140)^2}{140 + 169} = 63 \text{ kcal mole}^{-1}$$

$$Q_{NH} = \frac{Q_N^2}{Q_N + D_{NH}}, \text{ from ref. 25, } D_{NH} = 81 \text{ kcal mole}^{-1}, \text{ so } Q_{NH} = \frac{(140)^2}{140 + 81} = 89 \text{ kcal mole}^{-1}$$

If "intermediate" chemisorption is chosen for NH_2, then

$$Q_{NH_2} = \frac{1}{2}\left[\frac{(0.6)^2(140)^2}{\frac{(0.6)(140)}{2} + 169} + \frac{(140)^2}{140 + 169} \right] \begin{array}{l} = 48 \text{ kcal mole}^{-1} \text{ if } n' = 2 \\ = 46 \text{ kcal mole}^{-1} \text{ if } n' = 1 \end{array}$$

Problem 6.7 Solution

For acetone, $(CH_3)_2$ CO $\left(\text{or}\quad \begin{array}{c} CH_3 \\ C = O \\ CH_3 \end{array}\right)$, the enthalpy of formation is -217.3 kJ/mole or -51.9

kcal/mole [25]; therefore, $D_{Total} = D_{(CH_3)_2\,CO} = \Sigma \Delta H_{f_i}^o \,(\text{atoms}) - \Delta H_f^o$ so, based on:

and from reference [25]:

$$
\begin{array}{rcl}
3\,(C \rightarrow C\cdot) & = & 3\,(171.3\ \text{kcal/mole}) \\
3\,(H_2 \rightarrow 2\,H\cdot) & = & (3)\,(2)\,(51.2\ \text{kcal/mole}) \\
\tfrac{1}{2}\,(O_2 \rightarrow 2\,O\cdot) & = & (0.5)\,(119.2\ \text{kcal/mole}) \\
\Delta H_f^o & = & -(-51.9\ \text{kcal/mole}) \\
\hline
D_{Total} & = & 938.0\ \text{kcal/mole}
\end{array}
$$

From Table 6.5, on Pt: $Q_C = 150$ kcal/mole and $Q_O = 85$ kcal/mole.

(a)   For on-top adsorption $\left(\eta^1\mu_1\right)$, use equation 6.35 with n = 3 for Pt(111). Molecule will

bond O-end down; so

$Q_{oO} = (3/5)\,(85)$ and from Table 6.7b, the bond energy associated with each C-CH$_3$ group

is 376 kcal/mole, so

$D_{AB} = D_{C=O} = 938.0 - 2\,(376) = 186$ kcal/mole.

Then, for "weak" chemisorption, use equation 6.41:

$$
Q_{(CH_3)_2 CO} = \frac{Q_{oO}^2}{Q_{oO} + D_{(CH_3)_2 CO}} = \frac{(0.6)^2\,(85)^2}{(0.6)\,(85) + 186} = 11.0\ \text{kcal/mole}
$$

(b)   For di-σ-adsorption, i.e.,

$$\begin{array}{c} CH_3 \quad CH_3 \\ \diagdown C - O \diagup \\ Pt \rule{1.5cm}{0.4pt} Pt \end{array} \qquad \left(\eta^2 \mu_2\right)$$

equation 6.42 must be used, keeping in mind that atoms A and B can have other groups attached to alter the A-B bond energy, thus from part (a) $D_{AB} = D_{C=O} = 186$ kcal/mole.

From Eq. 6.43, $a = \dfrac{Q_{oC'}^2 \left(Q_{oC'} + 2\,Q_{oO}\right)}{\left(Q_{oC'} + Q_{oO}\right)^2}$ where $Q_{oC'}$ represents the heat of

chemisorption of the quasi-atomic $(CH_3)_2\, C\cdot$ fragment. For this relatively large species, a reasonable way to calculate its heat of chemisorption is to again use eq. 6.41 where D_{AB} now represents the total bond energy of all the bonds formed by the secondary C atom, thus from Table 6.7b:

$$D_{AB} = 2\left[D_{C-CH_3} - D_{CH_3}\right] = 2\left[376 - 293\right] = 166 \text{ kcal/mole}$$

so $Q_{oC'} = \dfrac{\left[(0.6)^2 \left((150)^2\right)\right]}{(0.6)\,(150) + 166} = 31.6$ kcal/mole, then

$$a = \frac{(31.6)^2 \left[31.6 + 2\,(51.0)\right]}{(31.6 + 51.0)^2} = \frac{1.33 \times 10^5}{6.82 \times 10^3} = 19.6 \text{ kcal/mole, and}$$

from Eq. 6.44, $\quad b = \dfrac{(51.0)^2 \left[51.0 + 2\,(31.6)\right]}{(51.0 + 31.6)^2} = \dfrac{2.97 \times 10^5}{6.82 \times 10^3} = 43.5$ kcal/mole

thus $\quad Q_{(CH_3)_2 CO} = \dfrac{ab\,(a + b) + D_{AB}\,(a - b)^2}{ab + D_{AB}\,(a + b)} \quad =$

$$\frac{(19.6)\,(43.5)\,(19.6 + 43.5) + (186)\,(19.6 - 43.5)^2}{(19.6)\,(43.5) + (186)\,(19.6 + 43.5)} = \frac{1.60 \times 10^5}{1.26 \times 10^4} = 12.7 \cong 13 \text{ kcal/mole}$$

Problem 6.8 Solution

Rate constant for bimolecular surface rxn or desorption $= A_b < 10^{-4} \ cm^2/s$

a.   Collision theory,   2-dimensional gas $2 \ A_{ad} \ \rightarrow \ A_{2 \ (g)}$

       $r = P \ Z_{AA} \ [A_{ad}]^2 = A_b \ [A_{ad}]^2$,   Probability P=1 ,   # collisions $= Z_{AA}$

       $Z_{AA} = \sigma \bar{v} = \sigma \left(\dfrac{8 \ k_B T}{\pi \ m} \right)^{1/2}$ where σ = molecular diameter (molecule sweeps out an area)

       $Z_{AA} \cong (2 \times 10^{-8} \ cm)(5 \times 10^4 \ cm/s) \simeq 10^{-3} \ cm^2/s$ (given in ref. 11, p. 68)

b.   Absolute rate theory, immobile adsorption: $\underbrace{2 \ A\text{-}S}_{\text{site pair}} \rightleftarrows S\text{-}A_2 - S \rightarrow A_2 + 2 \ S$

       $r = L_p \left(k_B T/h \right) \dfrac{Q^{\ddagger}}{Q^2_{A\text{-}S}} \left[A_{ad} \right]^2 = L_p \left(k_B T/h \right) \dfrac{Q^{\ddagger}}{Q_{SS}} \left[A_{ad} \right]^2$ where

       L_p = density of site pairs $= Z/2 \ L \ \Rightarrow \ (Z \cong 4)$,   then

       $A_b = \dfrac{Z}{2 \ L} \left(k_B T/h \right) \dfrac{Q^{\ddagger} \ (\approx 1)}{Q^2_{A\text{-}S} (\approx 1)} \cong \left(\dfrac{2}{10^{15}/cm^2} \right) (10^{13} \ s^{-1}) \cong 2 \times 10^{-2} \ cm^2/s$

c.   Absolute rate theory,   2-dimensional gas: $2 \ A_{ad} \ \rightleftharpoons \ A_{2 \ ad} \ \rightarrow \ A_{2,g}$

       $r = \left(k_B T/h \right) \dfrac{Q^{\ddagger}_{A_{2}ad}}{Q^2_{A_{ad}}} [A]^2 = \left(k_B T/h \right) \dfrac{q_{tr} \ (2 - D) \ q_{rot}}{q_{tr} \ (2 - D)^2} [A]^2 \ \Rightarrow \ A_b \cong \left(k_B T/h \right) \dfrac{(10^{17})(10)}{(10^{17})^2} \cong 10^{-3}$

Problem 6.9 Solution

Figure a. Curves for Activation Energy, Enthalpy, Entropy Determination
on activated carbon

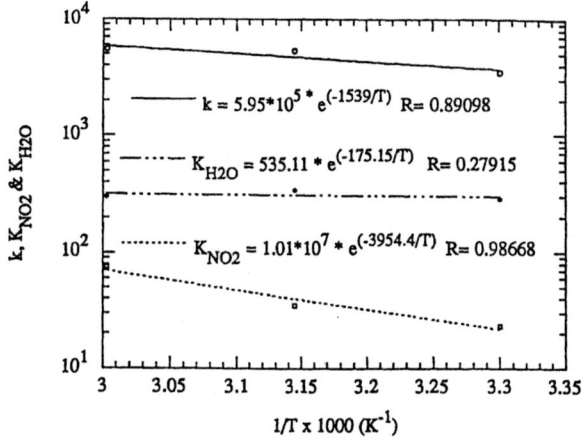

From slope & intercept:

1)   For NO$_2$

   $\Delta H^o_{ad} = + 7.9$ kcal mole^{-1}

   $\Delta S^o_{ad} = + 32.1$ e.u.   (1 e.u. = 1 cal/mole · K)

2)   For H$_2$O

   $\Delta H^o_{ad} = + 0.4$ kcal mole^{-1}

   $\Delta S^o_{ad} = + 12.5$ e.u.   Not consistent $\Rightarrow$ both ΔH & ΔS values are positive

Figure b. Curves for Activation Energy, Enthalpy and Entropy Determination on SiO_2

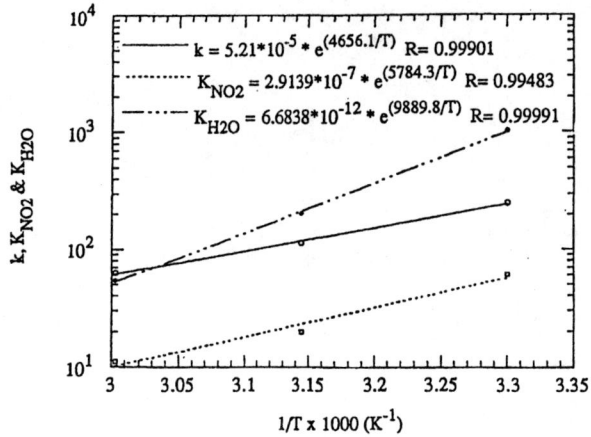

1) $E_a = -9.3$ kcal mole^{-1} $\Rightarrow$ negative and inconsistent

2)   <u>For NO_2</u>

$\Delta H^o_{ad} = -11.5$ kcal mole^{-1}

$\Delta S^o_{ad} = -29.9$ e.u.

$S^o_g = 57$ e.u.

3)   <u>For H_2O</u>

$\Delta H^o_{ad} = -19.7$ kcal mole^{-1}

$\Delta S^o_{ad} = -51.1$ e.u. $\Rightarrow$ inconsistent

$S^o_g = 45$ e.u. $< |-51.1$ e.u.$|$

Problem 7.1 Solution

(a) SSA states that $d[O]/dt = 0 \Rightarrow r_1 = r_2 \quad \therefore r = -d[O_3]/dt$

$$\frac{-d[O_3]}{dt} = r = 2\, r_1 = 2\, k_1\, [O_3][M] \quad \text{or} \quad \frac{d[O_2]}{dt} = 3\, k_1\, [O_3][M]$$

(b) Cannot assume Quasi-Equilibrated Adsorption (No Langmuir isotherm)

$r = \dfrac{d[CO_2]}{dt}$; @ steady state, so $r_1 = r_2$, Site Balance: $L = [S-O] + [S]$

$$k_1[N_2O][S] = k_2\,[CO][S-O] \Rightarrow [S-O] = (k_1/k_2)[N_2O][S]/[CO]$$

$$r = k_2\,[CO][S-O] = Lk_1 k_2\,[CO][N_2O]/(k_1\,[N_2O] + k_2\,[CO])$$

(c) $r = d[CH_4]/dt = k_2\,[CH_{2\,ad}][H_2]$ & $K = \dfrac{[C_2H_{2\,ad}][H_2]^2}{[C_2H_6][S]}$

I. Literature solution for reaction as written

If $[C_2H_{2\,ad}]$ is MARI, then $L = [C_2H_{2\,ad}] + [S]$ &

$$[C_2H_{2\,ad}] = K\,[C_2H_6][S]/[H_2]^2 = \frac{K[C_2H_6]}{[H_2]^2}\left(L - [C_2H_{2\,ad}]\right)$$

$$[C_2H_{2\,ad}]\left(1 + K[C_2H_6]/[H_2]^2\right) = \frac{LK[C_2H_6]}{[H_2]^2} \Rightarrow [C_2H_{2\,ad}] = \frac{LK[C_2H_6]}{[H_2]^2\left(1 + K[C_2H_6]/[H_2]^2\right)}$$

@ Steady State $\dfrac{d[CH_{2\,ad}]}{dt} = 0$: $\dfrac{d\varepsilon}{dt} = \dfrac{1}{2}\dfrac{d[CH_{2\,ad}]}{dt} = k_1\,[C_2H_{2\,ad}][H_2]$

$$d\varepsilon/dt = -\frac{d\left[CH_{2\,ad}\right]}{dt} = k_2\left[CH_{2\,ad}\right]\left[H_2\right] \Rightarrow 2r_1 = r_2$$

$$2\,k_1\left[C_2H_{2\,ad}\right]\left[H_2\right] = k_2\left[CH_{2\,ad}\right]\left[H_2\right] \Rightarrow \left[CH_{2\,ad}\right] = 2\left(k_1/k_2\right)\left[C_2H_2\right]_{ad}$$

$$d\left[CH_4\right]/dt = k_2\left[CH_{2\,ad}\right]\left[H_2\right] = \frac{2Lk_1K\left[C_2H_6\right]\left[H_2\right]}{\left[H_2\right]^2 + K\left[C_2H_6\right]} = \frac{2Lk_1K\left[C_2H_6\right]}{\left[H_2\right] + K\frac{\left[C_2H_6\right]}{\left[H_2\right]}}$$

II. Alternate solution for reaction written with well defined sites, S, i.e.,

(1) $C_2H_{6\,(g)} + S \overset{K}{\rightleftharpoons} C_2H_2\text{-}S + 2\,H_{2\,(g)}$

(2) $C_2H_2\text{-}S + H_{2\,(g)} + S \xrightarrow{\ k_1\ } 2\,CH_2\text{-}S$

(3) $\underline{2\,[CH_2\text{-}S + H_{2\,(g)} \xrightarrow{\ k_2\ } CH_{4\,(g)}]}$

$\qquad\qquad C_2H_6 + H_2 \Longrightarrow 2\,CH_4$

Note that a CH_2 group almost certainly requires a site for itself as shown in step 2.

$$r = \frac{d\left[CH_4\right]}{dt} = k_2\left[C_2H_2 - S\right]\left[H_2\right] \quad \& \quad K = \frac{\left[C_2H_2 - S\right]\left[H_2\right]^2}{\left[C_2H_6\right]\left[S\right]}$$

If $[C_2H_2 - S]$ is the MARI, then $L = [C_2H_2 - S] + [S]$ for the site balance, and

$$\left[C_2H_2 - S\right] = K\left[C_2H_6\right]\left[S\right]/\left[H_2\right]^2 = \left(K\left[C_2H_6\right]/\left[H_2\right]^2\right)\left(L - \left[C_2H_2 - S\right]\right) \Rightarrow$$

$$\left[C_2H_2 - S\right]\left(1 + K\left[C_2H_6\right]/\left[H_2\right]^2\right) = LK\left[C_2H_6\right]/\left[H_2\right]^2 \quad \&$$

$$\left[C_2H_2 - S\right] = \frac{LK\left[C_2H_6\right]}{\left[H_2\right]^2\left(1 + K\left[C_2H_6\right]/\left[H_2\right]^2\right)}$$

@ Steady state: $\dfrac{d\left[CH_2 - S\right]}{dt} = 0 \quad \& \quad \dfrac{d\varepsilon}{dt} = \dfrac{1}{2}\dfrac{d\left[CH_2 - S\right]}{dt} = k_1\left[C_2H_2 - S\right]\left[H_2\right]\left[S\right]\&$

$$\frac{d\varepsilon}{dt} = \frac{-d\left[CH_2 - S\right]}{dt} = k_2\left[CH_2 - S\right]\left[H_2\right] \Rightarrow 2\,r_1 = r_2, \text{ so}$$

$$2\,k_1\,[C_2H_2-S][H_2][S] = k_2\,[CH_2-S][H_2] \Rightarrow [CH_2-S] = 2\,(k_1/k_2)[C_2H_2-S][S]$$

$$[S] = L[H_2]^2 / ([H_2]^2 + K[C_2H_6]) \qquad \text{so}$$

$$\frac{d\,[CH_4]}{dt} = k_2[CH_2-S][H_2] = \left(\frac{Z}{2L}\right) 2L^2 k_1 K\,[C_2H_6][H_2]^3 / ([H_2]^2 + K\,[C_2H_6])^2$$

$$= \frac{k'\,[C_2H_6][H_2]^3}{([H_2]^2 + K\,[C_2H_6])^2} = \frac{k'\,[C_2H_6]}{[H_2]\left(1 + \dfrac{K\,[C_2H_6]}{[H_2]^2}\right)^2} = r$$

Problem 7.2 Solution

$$r = -\frac{d[A]}{dt} = \frac{d[B]}{dt} = k_3[B-S]$$

$$K_1 = \frac{[A-S]}{[A][S]} \quad , \quad K_2 = \frac{[B-S]}{[A-S]}$$

Site balance: $L = [S] + [A-S] + [B-S]$

(a) If $[B-S]$ is the MARI, then $L = [S] + [B-S]$ and

$$L = [S] + K_2[A-S] = [S] + K_1K_2[A][S] \quad \Rightarrow \quad [S] = \frac{L}{(1+K_1K_2[A])}$$

$$r = k_3K_2[A-S] = K_1K_2k_3[A][S] = \frac{Lk_3K_1K_2[A]}{(1+K_1K_2[A])}$$

(b) If $[A-S]$ is the MARI, then $L = [S] + [A-S]$ and $L = [S] + K_1[A][S] \quad \Rightarrow \quad [S] = \frac{L}{(1+K_1[A])}$

$$r = k_3K_2[A-S] = K_1K_2k_3[A][S] = \frac{Lk_3K_1K_2[A]}{(1+K_1[A])}$$

The mathematical forms are identical.

Problem 7.3 Solution

Write balanced sequence:

(1) $\qquad$ $4\ [NO + * \underset{}{\overset{K_{NO}}{\rightleftharpoons}} NO*]$

(2) $\qquad$ $NO* + * \underset{}{\overset{K_1}{\rightleftharpoons}} N* + O*$

(3) $\qquad$ $CH_4 + O* \underset{}{\overset{K_2}{\rightleftharpoons}} CH_2* + H_2O$

(4) $\qquad$ $CH_2* + NO* \xrightarrow{k} CHNO* + H*$

(5) $\quad H* + CHNO* + 2\ NO* \underset{}{\overset{K_3}{\rightleftharpoons}} CO_2 + N_2 + H_2O + N* + 3*$

(6) $\qquad$ $2\ N* \underset{}{\overset{1/K_{N_2}}{\rightleftharpoons}} N_2 + 2*$

$$4\ NO + CH_4 \implies 2\ N_2 + CO_2 + 2\ H_2O$$

Step 4 is RDS $\qquad \therefore\ r = k\ [CH_2*][NO*]$

$$[CH_2*] = \frac{K_2 P_{CH_4}[O*]}{P_{H_2O}} \quad , \quad [O*] = \frac{K_1[NO*][*]}{[N*]} \quad , \quad [NO*] = K_{NO} P_{NO}[*]$$

$$[O*] = \frac{K_1 K_{NO} P_{NO}[*]^2}{[N*]} \quad , \quad [N*]^2 = K_{N_2} P_{N_2}[*]^2$$

$$[O*] = \frac{K_1 K_{NO} P_{NO}[*]}{K_{N_2}^{1/2} P_{N_2}^{1/2}} \quad , \quad [CH_2*] = \frac{K_2 P_{CH_4} K_1 K_{NO}[*]}{K_{N_2}^{1/2} P_{N_2}^{1/2} P_{H_2O}}$$

$$r = \frac{k K_1 K_2 K_{NO}^2 P_{CH_4} P_{NO}^2[*]^2}{K_{N_2}^{1/2} P_{N_2}^{1/2} P_{H_2O}}$$

$$L = [*] + [NO*] + [CH_2*] = [*] + K_{NO} P_{NO}[*] + \frac{K_1 K_2 K_{NO} P_{CH_4} P_{NO}[*]}{K_{N_2}^{1/2} P_{N_2}^{1/2} P_{H_2O}}$$

$$[*] = \frac{L}{\left(1 + K_{NO} P_{NO} + \dfrac{K P_{CH_4} P_{NO}}{P_{N_2}^{1/2} P_{H_2O}}\right)} \qquad \text{where}\quad K = \frac{K_1 K_2 K_{NO}}{K_{N_2}^{1/2}}$$

and
$$r = \frac{LkK_1K_2K_{NO}^2P_{CH_4}P_{NO}^2}{K_{N_2}^{1/2}P_{N_2}^{1/2}P_{H_2O}\left(1+K_{NO}P_{NO}+\dfrac{KP_{CH_4}P_{NO}}{P_{N_2}^{1/2}P_{H_2O}}\right)^2}$$

Problem 7.4 Solution

$$r = \frac{-d[H_2O]}{dt} = \frac{d[H_2]}{dt} = \frac{d[CO_2]}{dt}$$

$$r = k_3[H_2O*] = k_4[O*][CO*] \quad \text{and} \quad K_1 = \frac{[CO*]}{P_{CO}[*]}$$

$$L = [*] + [CO*] + [H_2O*] + [O*] \qquad \text{(Site balance)}$$

Steady-state approximations:

$$\text{on } H_2O* \quad : \quad k_2 P_{H_2O}[*] = k_3[H_2O*]$$

$$\text{on } O* \quad : \quad k_3[H_2O*] = k_4[O*][CO*]$$

$$[H_2O*] = \left(\frac{k_2}{k_3}\right) P_{H_2O}[*] \quad \& \quad [O*] = \left(\frac{k_3}{k_4}\right)[H_2O*]/[CO*]$$

$$L = K_1 P_{CO}[*] + \left(\frac{k_2}{k_3}\right) P_{H_2O}[*] + [*] + \left(\frac{k_3}{k_4}\right)[H_2O*]/[CO*], \text{ the last term is negligible, so}$$

$$L = \left(K_1 P_{CO} + \left(\frac{k_2}{k_3}\right) P_{H_2O} + 1\right)[*]$$

$$r = k_3[H_2O*] = \frac{k_3 \cdot k_2 P_{H_2O}}{k_3}[*] = \frac{L k_2 P_{H_2O}}{\left(1 + K_1 P_{CO} + k' P_{H_2O}\right)}$$

Problem 7.5 Solution

(a)

$$A + S \underset{k_{-1}}{\overset{k_1}{\rightleftharpoons}} A \cdot S \qquad \text{reversible RDS}$$

$$B + S \overset{K_2}{\rightleftharpoons} B \cdot S$$

$$A \cdot S + B \cdot S \overset{K_3}{\rightleftharpoons} C \cdot S + D \cdot S$$

$$C \cdot S \overset{K_4}{\rightleftharpoons} C + S$$

$$D \cdot S \overset{K_5}{\rightleftharpoons} D + S$$

$$A + B \rightleftharpoons C + D$$

$r = k_1 P_A [S] - k_{-1}[A \cdot S]$, $L = [S] + [A \cdot S] + [B \cdot S] + [C \cdot S] + [D \cdot S]$ (from site balance)

$$[B \cdot S] = K_2 P_B [S] \ , \ [C \cdot S] = \frac{P_C [S]}{K_4} \ , \ [D \cdot S] = \frac{P_D [S]}{K_5} \ , \ [A \cdot S] = \frac{[C \cdot S][D \cdot S]}{K_3 [B \cdot S]}$$

$$K = \frac{P_C P_D}{P_A P_B} = K_1 K_2 K_3 K_4 K_5$$

$$[A \cdot S] = \frac{P_C P_D [S]}{K_4 K_5 K_3 K_2 P_B}$$

$$L = [S] \left(1 + P_C P_D \big/ K_2 K_3 K_4 K_5 P_B + K_2 P_B + \frac{1}{K_4 P_C} + \frac{1}{K_5 P_D} \right) \quad \Rightarrow$$

$$r = \left[L k_1 P_A - \frac{L k_{-1} P_C P_D}{K_2 K_3 K_4 K_5 P_B} \right] [S] = \frac{L k_1 P_A - L k_{-1} P_C P_D / K_2 K_3 K_4 K_5 P_B}{\left(1 + P_C P_D / K_2 K_3 K_4 K_5 P_B + K_2 P_B + P_C / K_4 + P_D / K_5 \right)}$$

or

Let $k_1 = k_A$, $k_{-1} = k_{-A}$, $K_2 = K_B$, $K_4 = 1/K_C$, $K_5 = 1/K_D$, then

$$K_3 = K/K_1 K_2 K_4 K_5 = K K_C K_D / K_A K_B \quad \text{and} \quad K_A = k_A/k_{-A} \quad \text{so,}$$

$$K_2 K_3 K_4 K_5 = K/K_1 = K/K_A = K k_{-A}/k_A = K k_{-1}/k_1$$

substitution gives

$$r = \frac{L k_1 \left(P_A - P_C P_D / K\, P_B \right)}{\left(1 + K_A\, P_C P_D / K\, P_B + K_B P_B + K_C P_C + K_D P_D \right)}$$

(b) $\qquad A + S \; \overset{K_1}{=\!\!\ominus\!\!=} \; A \cdot S$

$$B + S \; \overset{K_2}{=\!\!\ominus\!\!=} \; B \cdot S$$

$$A \cdot S + B \cdot S \; \overset{K_3}{=\!\!\ominus\!\!=} \; C \cdot S + D \cdot S$$

$$C \cdot S \; \overset{K_4}{=\!\!\ominus\!\!=} \; C + S$$

$$D \cdot S \; \overset{k_5}{\underset{k_{-5}}{\rightleftharpoons}} \; D + S \qquad\qquad \text{Reversible RDS}$$

_____

$$A + B \; \rightleftharpoons \; C + D$$

$r = k_5 [D \cdot S] - k_{-5} P_D [S]$, $L = [A \cdot S] + [B \cdot S] + [C \cdot S] + [D \cdot S] + [S]$ (from site balance)

$$[A \cdot S] = K_1 P_A [S], \; [B \cdot S] = K_2 P_B [S], \; [C \cdot S] = P_C [S] / K_4 \; , \; K_3 = \frac{[C \cdot S][D \cdot S]}{[A \cdot S][B \cdot S]} \quad \text{so}$$

$$[D \cdot S] = K_3 [A \cdot S][B \cdot S] / [C \cdot S] = K_1 K_2 K_3 K_4 P_A P_B [S] / P_C$$

$$L = [S] \left(1 + K_1 P_A + K_2 P_B + P_C / K_4 + K_1 K_2 K_3 K_4 P_A P_B / P_C \right) \qquad \text{so}$$

$$r = \frac{Lk_5 K_1 K_2 K_3 K_4 P_A\,P_B/P_C - Lk_{-5}P_D}{\left(1 + K_1 P_A + K_2 P_B + P_C/K_4 + K_1 K_2 K_3 K_4 P_A\,P_B/P_C\right)}$$

Now $K = K_1 K_2 K_3 K_4 K_5$ and $K_1 = K_A$, $K_2 = K_B$, $1/K_4 = K_C$,

$$1/K_5 = K_D = k_{-5}/k_5\ ,\ K_1 K_2 K_3 K_4 = K/K_5 = k_{-5}\,K/k_5$$

substitution gives

$$r = \frac{Lk_{-5}KP_A\,P_B/P_C - Lk_{-5}P_D}{\left(1 + K_A P_A + K_B P_B + K_C P_C + K_D K P_A\,P_B/P_C\right)}$$

(c) $A_2 + 2S \underset{k_{-1}}{\overset{k_1}{\rightleftharpoons}} 2\,A\cdot S$ reversible RDS

$2\,[B + S \overset{K_2}{\rightleftharpoons} B\cdot S]$

$2\,[A\cdot S + B\cdot S \overset{K_3}{\rightleftharpoons} C\cdot S + S]$

$\underline{2\,[C\cdot S \overset{K_4}{\rightleftharpoons} C + S]}$

$A_2 + 2B \rightleftharpoons 2C$

$r = k_1 P_{A_2}[S]^2 - k_{-1}[A\cdot S]^2$, $L = [A\cdot S] + [B\cdot S] + [C\cdot S] + [S]$ (from site balance)

$$[B\cdot S] = K_2 P_B [S]\ ,\ \ [C\cdot S] = \frac{P_C[S]}{K_4}\ ,\ \ K_3 = \frac{[C\cdot S][S]}{[A\cdot S][B\cdot S]}$$

$$[A\cdot S] = \frac{[C\cdot S][S]}{K_3[B\cdot S]}\ ,\ \ K = K_1 K_2^2 K_3^2 K_4^2\ \text{ or }\ K^{1/2} = K_1^{1/2} K_2 K_3 K_4\ ,\ \ \text{so}$$

$$[A\cdot S] = \frac{P_C[S]}{K_2 K_3 K_4 P_B}$$

$$L = |S|\left(1 + P_C/K_2K_3K_4P_B + K_2P_B + P_C/K_4\right) \quad \text{so}$$

$$r = \frac{(Z/2L)L^2k_1P_{A_2} - (Z/2L)L^2k_{-1}\left(P_C/K_2K_3K_4P_B\right)^2}{\left(1 + P_C/K_2K_3K_4P_B + K_2P_B + P_C/K_4\right)^2}$$

with site-pair probability of $(Z/2L)$

where Z = coordination number

$$\text{or, because}$$

$$K_2K_3K_4 = K^{1/2}/K_1^{1/2}$$

$$r = \frac{Lk_1'\left(P_{A_2} - P_C^2/KP_B^2\right)}{\left(1 + K_1^{1/2}P_C/K^{1/2}P_B + K_BP_B + K_CP_C\right)^2}$$

(d)
$$A_2 + 2S \overset{K_1}{\rightleftharpoons} 2A \cdot S$$

$$2\,[B + S \overset{K_2}{\rightleftharpoons} B \cdot S]$$

$$2\,[A \cdot S + B \cdot S \overset{K_3}{\rightleftharpoons} C \cdot S + S]$$

$$2\,[C \cdot S \underset{k_{-4}}{\overset{k_4}{\rightleftharpoons}} C + S] \qquad \text{(reversible RDS)}$$

$$\overline{\qquad\qquad\qquad\qquad\qquad\qquad}$$

$$A_2 + 2B \rightleftharpoons 2C$$

$$r = k_4[C \cdot S] - k_{-4}P_C[S] \quad, \quad [A \cdot S]^2 = K_1P_{A_2}[S]^2 \quad, \quad [B \cdot S] = K_2P_B[S] \quad,$$

$$K_3 = \frac{[C \cdot S][S]}{[A \cdot S][B \cdot S]} \Rightarrow [C \cdot S] = \frac{K_3[A \cdot S][B \cdot S]}{[S]} = K_1^{1/2}K_2K_3P_{A_2}^{1/2}P_B[S]$$

$$L = [S] + [A \cdot S] + [B \cdot S] + [C \cdot S] = [S]\left(1 + K_1^{1/2}P_{A_2}^{1/2} + K_2P_B + K_1^{1/2}K_2K_3P_{A_2}^{1/2}P_B\right) \quad \text{(from site balance)}$$

$$K = K_1K_2^2K_3^2K_4^2 \Rightarrow K_1^{1/2}K_2K_3 = K^{1/2}/K_4 \qquad \text{so}$$

$$r = \frac{(Z/2L)L^2 k_4 K_1^{1/2} K_2 K_3 P_{A_2}^{1/2} P_B - (Z/2L)L^2 k_{-4} P_C}{\left(1 + K_1^{1/2} P_{A_2}^{1/2} + K_2 P_B + K_1^{1/2} K_2 K_3 P_{A_2}^{1/2} P_B\right)}$$

Now with $\quad K_1 = K_{A_2} \quad , \quad K_2 = K_B \quad , \quad 1/K_4 = K_C \quad \Rightarrow \quad K = K_A K_B^2 K_3^2 K_4^2 \quad$ so

$$r = \frac{L(Z/2)k_{-4}\left(K^{1/2} P_{A_2}^{1/2} P_B - P_C\right)}{\left(1 + K_{A_2}^{1/2} P_{A_2}^{1/2} + K_B P_B + K K_C P_{A_s}^{1/2} P_B\right)}$$

Problem 7.6 Solution

(a) $\quad r = \dfrac{-d\lfloor C_2H_6\rfloor}{dt} = Lk_3\theta_{C_2H_4}P_{H_2}$

$$K_2 = \dfrac{\theta_{C_2H_4}P_{H_2}}{\theta_{C_2H_5}\theta_H}$$

Use steady-state approximation for total surface carbon atoms to get another equation for $\theta_{C_2H_5}$,

i.e., $\dfrac{d\theta_{C\,(total)}}{dt} = 0$. Then, because steps 2 and 4 are very rapid:

$$\dfrac{d\theta_{C\,Total}}{dt} = k_1P_{C_2H_6} - Lk_{-1}\theta_{C_2H_5}\theta_H - Lk_3\theta_{C_2H_4}P_{H_2} = 0 \;\Rightarrow$$

$$k_1P_{C_2H_6} - Lk_3\theta_{C_2H_4}P_{H_2} = Lk_{-1}\theta_{C_2H_5}\theta_H \;\Rightarrow\; \theta_{C_2H_5} = \dfrac{k_1P_{C_2H_6} - Lk_3\theta_{C_2H_4}P_{H_2}}{Lk_{-1}\theta_H}$$

$$\theta_{C_2H_4} = \left(\dfrac{K_2\theta_H}{P_{H_2}}\right)\left(\dfrac{k_1P_{C_2H_6} - Lk_3\theta_{C_2H_4}P_{H_2}}{Lk_{-1}\theta_H}\right) = \dfrac{K_2k_1P_{C_2H_6}}{Lk_{-1}P_{H_2}} - \dfrac{LK_2k_3\theta_{C_2H_4}}{Lk_{-1}} \;\Rightarrow$$

$$\theta_{C_2H_4}\left(1 + K_2k_3/k_{-1}\right) = \dfrac{k_1K_2P_{C_2H_6}}{Lk_{-1}P_{H_2}} \;\Rightarrow\; \theta_{C_2H_4} = \dfrac{\left(k_1K_2/Lk_{-1}\right)P_{C_2H_6}/P_{H_2}}{\left(1 + K_2k_3/k_{-1}\right)} \qquad \text{and}$$

$$r = \dfrac{Lk_3P_{H_2}k_1K_2\,P_{C_2H_6}/Lk_{-1}P_{H_2}}{\left(1 + K_2\,k_3/k_{-1}\right)} = \dfrac{k_1K_2k_3P_{C_2H_6}}{\left(1 + K_2\,k_3/k_{-1}\right)} = k'P_{C_2H_6}$$

but $\theta_{C_2H_4}$ can be rewritten to give:

$$\theta_{C_2H_4} = \dfrac{\left(k_1/k_3\right)P_{C_2H_6}/P_{H_2}}{L\left(1 + k_{-1}/K_2k_3\right)} \qquad \text{so that} \qquad r = \dfrac{k_1P_{C_2H_6}}{1 + k_{-1}/K_2k_3}$$

which is notation consistent with reference 55.

(b)

$$C_2H_6 + 2S \xrightleftharpoons[k_{-1}]{k_1} C_2H_5 \cdot S + H \cdot S$$

$$C_2H_5 \cdot S + H \cdot S \underset{\ominus}{\overset{K_2}{\rightleftharpoons}} C_2H_4 \cdot S + S + H_2$$

$$C_2H_4 \cdot S + H_2 + S \xrightarrow{k_3} 2 CH_3 \cdot S$$

$$\underline{2 CH_3 \cdot S + H_2 \underset{\ominus}{\overset{K_4}{\rightleftharpoons}} 2 CH_4 + 2S}$$

$$C_2H_6 + H_2 \implies 2 CH_4$$

$$r = -\frac{d[C_2H_6]}{dt} = k_3[C_2H_4 \cdot S][S]P_{H_2}$$

$$K_2 = \frac{[C_2H_4 \cdot S][S]P_{H_2}}{[C_2H_5 \cdot S][H \cdot S]}$$

Steady-state approximation on all surface C atoms gives:

$$k_1 P_{C_2H_6}[S]^2 - k_{-1}[C_2H_5 \cdot S][H \cdot S] - k_3[C_2H_4 \cdot S][S]P_{H_2} = 0$$

$$k_1 P_{C_2H_6}[S]^2 - k_3[C_2H_4 \cdot S][S]P_{H_2} = k_{-1}[C_2H_5 \cdot S][H \cdot S] \quad \text{and}$$

$$[C_2H_5 \cdot S] = \frac{k_1 P_{C_2H_6}[S]^2}{k_{-1}[H \cdot S]} - \frac{k_3[C_2H_4 \cdot S][S]P_{H_2}}{k_{-1}[H \cdot S]}$$

$$[C_2H_4 \cdot S] = \left(\frac{K_2[H \cdot S]}{[S]P_{H_2}}\right)\left(\frac{k_1 P_{C_2H_6}[S]^2 - k_3[C_2H_4 \cdot S][S]P_{H_2}}{k_{-1}[H \cdot S]}\right)$$

$$[C_2H_4 \cdot S](1 + K_2 k_3/k_{-1}) = \frac{k_1 K_2 P_{C2H6}}{k_{-1}P_{H_2}} \Rightarrow [C_2H_4 \cdot S] = \frac{(k_1 K_2/k_{-1})P_{C2H6}/P_{H_2}}{1 + K_2 k_3/k_{-1}}$$

Now, site balance to get [S] is:

$$L = [S] + [H \cdot S] + [C_2H_5 \cdot S] + [C_2H_4 \cdot S] + [CH_3 \cdot S]$$

Agreement with (a) is achieved only if $[S] \cong L$, i.e., the surface is essentially free of all adsorbed

species, i.e.,

$\theta_H, \theta_{C_2H_5}, \theta_{C_2H_4}, \theta_{CH_3} \ll 1$. This a questionable assumption.

Problem 7.7 Solution

Plotting ℓn rate vs. ℓn P_i using a power rate law gives the following reaction orders:

T(K)	Reaction Order		
	N_2O	O_2	N_2
623	0.08	-0.31	0
653	0.24	-0.12	0
673	0.31	-0.07	0

The simplest L-H model would be for unimolecular decomposition:

(1) $\qquad 2\,[N_2O + * \underset{}{\overset{K_{N_2O}}{\rightleftharpoons}} N_2O\,*]$

(2) $\qquad 2\,[N_2O\,* \xrightarrow{\;k\;} N_2 + O*]$

(3) $\qquad \underline{2\,O* \overset{1/K_{O_2}}{\rightleftharpoons} O_2 + 2\,*}$

$\qquad\qquad 2\,N_2O \implies 2\,N_2 + O_2$

$$r_m = \frac{1}{m}\frac{dN_{N_2}}{dt} = \frac{1}{m}\left(-\frac{dN_{N_2O}}{dt}\right) = k\left[N_2O\,*\right]$$

From (1): $K_{N_2O} = \dfrac{[N_2O*]}{P_{N_2O}[*]}$, so $\;[N_2O*] = K_{N_2O}P_{N_2O}[*]$

From (3): $K_{O_2} = \dfrac{[O*]^2}{P_{O_2}[*]^2}$, so $\;[O*] = K_{O_2}^{1/2}P_{O_2}^{1/2}[*]$

Site balance gives: $\quad L = [*] + [N_2O*] + [O*] \qquad$ thus

$$L = [*] + K_{N_2O}P_{N_2O}[*] + K_{O_2}^{1/2}P_{O_2}^{1/2}[*], \text{ and } [*] = L\Big/\Big(1 + K_{N_2O}P_{N_2O} + K_{O_2}^{1/2}P_{O_2}^{1/2}\Big),$$

consequently,

$$r = kK_{N_2O}P_{N_2O}[*] = LkK_{N_2O}\,P_{N_2O}\Big/\Big(1 + K_{N_2O}P_{N_2O} + K_{O_2}^{1/2}P_{O_2}^{1/2}\Big)$$

Arrhenius plots of the fitting parameters listed in Table 2 provide the following values:

For K_{N_2O} : $\Delta H^\circ_{ad} = -17$ kcal mole^{-1} and $\Delta S^\circ_{ad} - 21$ cal mole^{-1} K^{-1} (e.u.)

For K_{O_2} : $\Delta H^\circ_{ad} = -25$ kcal mole^{-1} and $\Delta S^\circ_{ad} = -35$ e.u.

For k : $E_{RDS} = 57$ kcal mole^{-1}

The enthalpy and entropy values for adsorption fulfill all the guidelines in Table 6.9, thus they are consistent.

From either a linear extrapolation of the high-P portions of the two isotherms in Figure 1 or using the difference between the two at 100 Torr CO pressure, the irreversible uptake is 580 μmole CO g^{-1}_{cat}. The dispersion of Cu is: $D_{Cu} = Cu_s / Cu_{tot}$, and with $CO_{ad}/Cu_s = 1$,

$$D_{Cu} = \frac{580 \; \mu\text{mole Cu}_s \; g^{-1}_{cat}}{\left(0.0456 \text{ g Cu } g^{-1}_{cat}\right)\left(\text{mole Cu}/63.55 \text{ g Cu}\right)\left(10^6 \; \mu\text{mole/mole}\right)} = 0.81$$

Under differential reaction conditions, $P_{O_2} \approx 0$ and can be ignored; therefore, an easy way is to choose a known differential rate and correct for temperature, for example:

$$\frac{r_{823K}}{r_{673K}} = \frac{k_{823K}}{k_{673K}} = \frac{e^{-36200 \; \text{cal/mole}/(1.987 \; \text{cal/mole·K})(823K)}}{e^{-36200 \; \text{cal/mole}/(1.987 \; \text{cal/mole·K})(673K)}} = 139 \quad \text{thus}$$

$$TOF_{823K} = \frac{r_{823K}}{Cu_s} = \frac{\left(12.6 \; \mu\text{mole/s·g}\right)\left(13 \text{ atm}^{-1}\right)\left(0.0666 \text{ atm}\right)(139)}{\left[1 + \left(13 \text{ atm}^{-1}\right)\left(0.0666 \text{ atm}\right)\right]\left(580 \; \mu\text{mole Cu}_s /g\right)} = 1.4 \; s^{-1}$$

Problem 7.8 Solution

Step 2 defines the rate:

$$r_m = \frac{1}{m}\frac{dN_{N_2}}{dt} = \frac{1}{m}\left(-\frac{dN_{N_2O}}{dt}\right) = k[N_2O*]$$

Step 1 gives: $K_{N_2O} = \dfrac{[N_2O*]}{P_{N_{2O}}[*]}$, so $[N_2O*] = K_{N_2O}P_{N_2O}[*]$

Assuming all surface species are included, a site balance gives

$$L = [*] + [N_2O*] + [O*]$$

To remove the unknown $[O*]$, the SSA must be used :

$$\frac{d[O*]}{dt} = k[N_2O*] + k_{-1}P_{O_2}[*]^2 - k_1[O*]^2 = 0$$

and

$$[O*] = \left(\frac{k[N_2O*] + k_{-1}P_{O_2}[*]^2}{k_1}\right)^{1/2}$$

Then $L = [*] + K_{N_2O}P_{N_2O}[*] + \left(\dfrac{k[N_2O*] + k_{-1}P_{O_2}[*]^2}{k_1}\right)^{1/2} = z[*] + \left(x[*] + y[*]^2\right)^{1/2}$

where $z = 1 + K_{N_2O}P_{N_2O}$, $x = kK_{N_2O}\,P_{N_2O}/k_1$, , and $y = k_{-1}\,P_{O_2}/k_1$.

Rearranging and squaring each side gives:

$$x[*] + y[*]^2 = (L - z[*])^2 = L^2 - 2Lz[*] + z^2[*]^2$$

and

$$(y - z^2)[*]^2 + (x + 2Lz)[*] - L^2 = 0$$

The solution for this quadratic expression is:

$$[*] = \frac{(x + 2Lz) \pm (x^2 + 4Lxz + 4yL^2)^{1/2}}{2(y - z^2)}$$

Substituting back and using the positive root gives:

$$[*] = \frac{kK_{N_2O}\,P_{N_2O}/k_1 + 2L(1 + K_{N_2O}P_{N_2O})}{-2\left[k_{-1}P_{O_2}/k_1 - (1 + K_{N_2O}P_{N_2O})^2\right]} +$$

$$\frac{\left[\left(\dfrac{kK_{N_2O}P_{N_2O}}{k_1}\right)^2 + \dfrac{4LkK_{N_2O}P_{N_2O}(1 + K_{N_2O}P_{N_2O})}{k_1} + \dfrac{4L^2k_{-1}P_{O_2}}{k_1}\right]^{1/2}}{-2\left[k_{-1}P_{O_2}/k_1 - (1 + K_{N_2O}P_{N_2O})^2\right]}$$

Because the rate is : $r_m = k[N_2O *] = kK_{N_2O}P_{N_2O}[*]$, the final rate expression is :

$$r_m = \frac{d\,P_{N_2O}\left\{1 + (a + c)\,P_{N_2O} + \left[a\,P_{N_2O} + (a^2 + ac)P_{N_2O} + b\,P_{O_2}\right]^{1/2}\right\}}{(1 + c\,P_{N_2O})^2 - b\,P_{O_2}}$$

where $a = kK_{N_2O}/k_1$, $b = k_1/k_{-1}$, $c = K_{N_2O}$ and $d = LkK_{N_2O}$. Note that these four

parameters can be combined in various ways to give the values in Table 1.

An Arrhenius plot of the adsorption equilibrium constant, K_{N_2O}, gives $\Delta H_{ad}^o = -25.2$

kcal mole^{-1} and $\Delta S_{ad}^o = -33$ e.u., which satisfy the guidelines in Table 6.9.

The rate constant Lk represents a unimolecular decomposition reaction on the surface, so

guideline 2 in Table 6.10 can be applied:

$$r_a = r_m / A_{cu,g} = \left(\frac{1.1\,\mu mole}{s \cdot g\,cat} \right) \left(\frac{1\,g\,cat}{2(354\,\mu mole\,Cu_s)} \right) \left(\frac{\sim 10^{15}\,Cu_s}{cm^2\,Cu} \right) \cong 2x10^{12}\,\frac{molecule}{s\,cm^2}$$

or, with correction for activation energy:

$$LA_d e^{-34400/1.987 \cdot 843} = 2x10^{12} = 1.2x10^{-9}\,LA_d \quad and \quad LA_d = 2x10^{21}\,molecule/s\,cm_s$$

Both values are below 10^{28} molecule/s cm^2.

Problem 7.9 Solution

One has dissociative H_2 adsorption, and to get a negative dependence on a reactant (acetone), the denominator must be squared, which indicates a bimolecular surface reaction as a RDS with competitive adsorption between acetone (Ace) molecules and H atoms for active sites, so:

(1) $\qquad$ $H_2 + 2\,S \xrightleftharpoons{K_{H_2}} 2\,H\text{-}S$

(2) $\qquad$ $Ace + S \xrightleftharpoons{K_{Ace}} Ace\text{-}S$

(3) $\qquad$ $Ace\text{-}S + H\text{-}S \xrightleftharpoons{K_3} AceH\text{-}S + S$

(4) $\qquad$ $AceH\text{-}S + H\text{-}S \xrightarrow{k_4} IPA\text{-}S + S$

(5) $\qquad$ $IPA\text{-}S \xrightleftharpoons{1/K_{IPA}} IPA + S$

$Ace + H_2 \implies IPA$ (Isopropyl alcohol)

To get a 1st-order dependence on H_2, step 4 and not step 3 must be the RDS.

The overall site balance appears to be simplified to:

$$L = [S] + [Ace] = [S] + K_{Ace} P_{Ace} [S] \qquad (1)$$

because there is no dependence on IPA at these differential conversions and the dependence on P_{H_2} must be as close to unity as possible, so $[Ace-S]$ is the MARI. Therefore,

$$r_m = \frac{1}{m}\frac{dN_{IPA}}{dt} = k_4 [AceH - S][H - S] \qquad (2)$$

and

$$[H-S] = K_{H_2}^{1/2} P_{H_2}^{1/2}[S] \quad \text{from step 1.} \tag{3}$$

Now steps 2 and 3 can be added to get

$$\text{(6)} \qquad \text{Ace} + \text{H-S} \xrightleftharpoons[]{K_{Ace}K_3} \text{AceH-S}$$

so

$$[AceH-S] = K_{Ace}K_3 P_{Ace}[H-S] = K_{H_2}^{1/2} K_{Ace} K_3 P_{H_2}^{1/2} P_{Ace} \tag{4}$$

Consequently, from equation 1:

$$[S] = L/(1 + K_{Ace}P_{Ace}) \tag{5}$$

and

$$r_m = k_4 K_{Ace}K_3 P_{Ace}[H-S]^2 = k_4 K_{Ace} K_{H_2} K_3 P_{Ace} P_{H_2}[S]^2 = \frac{Lk_4' K_{Ace} K_{H_2} K_3 P_{Ace} P_{H_2}}{(1 + K_{Ace}P_{Ace})^2} \tag{6}$$

$(Q_{ad} \rightarrow \quad \text{Add later})$

Problem 7.10 Solution

(a)   The rate, defined by N_2 formation, is $r = \dfrac{d[N_2]}{dt} = \dfrac{-1}{2}\dfrac{d[NO]}{dt} = k_2[N*]^2$

From the 3 quasi-equilibrated steps:

$$[NO*] = K_{NO}P_{NO}[*], \; [N*][O*] = K_1[NO*][*] \text{ and } [O*] = K_{O_2}^{1/2}P_{O_2}^{1/2}[*]$$

$$[N*] = \frac{K_1[NO*][*]}{[O*]} = \frac{K_1 K_{NO}P_{NO}[*]^2}{[O*]} = \frac{K_1 K_{NO}P_{NO}[*]^2}{K_{O_2}^{1/2}P_{O_2}^{1/2}[*]}$$

Site balance gives : $L = [*] + [NO*] + [N*] + [O*] \quad \Rightarrow$

$$L = [*] + K_{NO}P_{NO}[*] + \frac{K_1 K_{NO}P_{NO}}{K_{O_2}^{1/2}P_{O_2}^{1/2}}[*] + K_{O_2}^{1/2}P_{O_2}^{1/2}[*] \quad \text{so}$$

$$[*] = L \Big/ \Big(1 + K_{NO}P_{NO} + K_{O_2}^{1/2}P_{O_2}^{1/2} + K_1 K_{NO} P_{NO}/K_{O_2}^{1/2}P_{O_2}^{1/2}\Big),$$

for site pairs : $\dfrac{Z}{2L}$ is the probability factor, so

$$r = \frac{ZL}{2}k_2 K_1^2 K_{NO}^2 P_{NO}^2 [*]^2 / K_{O_2}P_{O_2}$$

$$= L k_2' K_1^2 K_{NO}^2 P_{NO}^2 / K_{O_2}P_{O_2} \Big(1 + K_{NO}P_{NO} + K_{O_2}^{1/2}P_{O_2}^{1/2} + K_1 K_{NO} P_{NO}/K_{O_2}^{1/2}P_{O_2}^{1/2}\Big)^2$$

or $r = k P_{NO} \Big/ \Big(K_{O_2}^{1/2}P_{O_2}^{1/2} + K_{O_2}^{1/2}K_{NO}P_{O_2}^{1/2}P_{NO} + K_{O_2}P_{O_2} + K_1 K_{NO}P_{NO}\Big)^2$

(b) $[NO*] \approx 0 \Rightarrow r = k P_{NO}^2 \Big/ \Big(K_{O_2}^{1/2}P_{O_2}^{1/2} + K_{O_2}P_{O_2} + K_1 K_{NO}P_{NO}\Big)^2$

(c) $[NO*]$ is MARI $\Rightarrow r = k P_{NO}^2 \Big/ \Big(K_{O_2}^{1/2}P_{O_2}^{1/2} + K_{O_2}^{1/2}K_{NO}P_{O_2}^{1/2}P_{NO}\Big)^2$

(d) $[O*] \approx 0 \Rightarrow r = k P_{NO}^2 \Big/ \Big(K_{O_2}^{1/2}P_{O_2}^{1/2} + K_{O_2}^{1/2}K_{NO}P_{O_2}^{1/2}P_{NO} + K_1 K_{NO}P_{NO}\Big)^2$

(e) $[O*]$ is MARI $\Rightarrow r = k P_{NO}^2 \Big/ \Big(K_{O_2}^{1/2}P_{O_2}^{1/2} + K_{O_2}P_{O_2}\Big)^2$

(f) $[N*]$ is MARI $\Rightarrow r = k\,P_{NO}^2 \Big/ \Big(K_{O_2}^{1/2}P_{O_2}^{1/2} + K_1 K_{NO}P_{NO}\Big)^2$

(g) $[N*]$ near saturation $(>> [*]) \Rightarrow [N*] \cong L \Rightarrow r = \dfrac{Z}{2L}k_2L^2 = L\,k_2'$

Reject: (e) – Cannot give reaction order on P_{NO} below 2

(g) – Observed reaction order is not zero order

Problem 7.11 Solution

The rate can be defined as $r_m = \dfrac{1}{m}\dfrac{dN_{N_2}}{dt} = \dfrac{1}{m}\left(-\dfrac{dN_{CH_4}}{dt}\right)$ for the reaction

$CH_4 + 2\,O_2 \rightarrow CO_2 + 2\,H_2O$. With dissociative O_2 adsorption:

(1) $\qquad O_2 + 2* \overset{K_{O_2}}{\rightleftharpoons} 2\,O*$

(2) $\qquad CH_4 + * \overset{K_{CH_4}}{\rightleftharpoons} CH_4*$

(3) $\qquad CH_4* + O* \overset{k_1}{\longrightarrow} CH_3* + OH*$

(4) $\qquad CH_3* + O* \overset{K_2}{\rightleftharpoons} CH_2O* + H*$

(5) $\qquad CH_2O* + O* \overset{K_3}{\rightleftharpoons} HCO* + OH*$

(6) $\qquad HCO* + O* \overset{K_4}{\rightleftharpoons} CO* + OH*$

(7) $\qquad H* + OH* \overset{K_5}{\rightleftharpoons} H_2O* + *$

(8) $\qquad CO* + O* \overset{K_6}{\rightleftharpoons} CO_2* + *$

(9) $\qquad CO_2* \overset{1/K_{CO_2}}{\rightleftharpoons} CO_2 + *$

(10) $\qquad 2\,OH* \overset{K_7}{\rightleftharpoons} H_2O* + O*$

(11) $\qquad 2\,[H_2O* \overset{1/K_{H_2O}}{\rightleftharpoons} H_2O + *]$

(12) $\qquad CO* \overset{1/K_{CO}}{\rightleftharpoons} CO + *$

$$CH_4 + 2\,O_2 \implies CO_2 + 2\,H_2O$$

Step 12 allows for CO desorption as a product. The rate based on this sequence is given by step 3 (the RDS):

$$r_m = k_1 [CH_4\,*][O\,*]$$

From step 1, $\quad K_{O_2} = [O\,*]^2/P_{O_2}[*]^2 \quad$ and $\quad [O\,*] = K_{O_2}^{1/2} P_{O_2}^{1/2}[*]$

From step 2, $\quad K_{CH_4} = [CH_4\,*]/P_{CH_4}[*] \quad$ and $\quad [CH_4\,*] = K_{CH_4} P_{CH_4}[*]$

Thus $\quad r_m = k_1 K_{CH_4} K_{O_2}^{1/2} P_{CH_4} P_{O_2}^{1/2}[*]^2$

The site balance with only adsorbed reactants and products is:

$$L = [CH_4\,*] + [O\,*] + [CO_2\,*] + [H_2O\,*] + [CO\,*] + [*]$$

Steps 9, 11, and 12 give, respectfully:

$$[CO_2\,*] = K_{CO_2} P_{CO_2}[*],\ [H_2O\,*] = K_{H_2O} P_{H_2O}[*],\ \text{and}\ [CO\,*] = K_{CO} P_{CO}[*]$$

thus $\quad [*] = L/\left(1 + K_{CH_4} P_{CH_4} + K_{O_2}^{1/2} P_{O_2}^{1/2} + K_{CO_2} P_{CO_2} + K_{H_2O} P_{H_2O} + K_{CO} P_{CO}\right)$

and

$$r_m = L k_1' K_{CH_4} K_{O_2}^{1/2} P_{CH_4}\ P_{O_2}^{1/2}/\left(1 + K_{CH_4} P_{CH_4} + K_{O_2}^{1/2} P_{O_2}^{1/2} + K_{CO_2} P_{CO_2} + K_{H_2O} P_{H_2O} + K_{CO} P_{CO}\right)^2$$

From Arrhenius plots of K_{CH_4} & K_{O_2} ,

For $CH_4 := \Delta H_{ad}^\circ = -20$ kcal mole^{-1}, $\Delta S_{ad}^\circ = -13$ e.u.;

For $O_2 : \Delta H_{ad}^\circ = -30$ kcal mole^{-1}, $\Delta S_{ad}^\circ = -25$ e.u. (1 e.u. = 1 cal mole^{-1} K^{-1})

Problem 7.12 Solution

(a) $\quad r_{mNO} = \dfrac{1}{m}\dfrac{dN_i}{v_i dt} = \dfrac{-1}{2m}\dfrac{dN_{NO}}{dt}$

$r_{mNO} = L_* L_s k_o \theta_{NO} \theta + L_* k_1 \theta_{NO} \theta_{HNO}$ where L_* and L_s are site densities of $*$ and S sites, respectively.

SSA on HNO$*$ gives: $\dfrac{d[HNO*]}{dt} = L_* L_s k_o \theta_{NO} \theta_H - L_* k_1 \theta_{NO} \theta_{HNO} = 0$ and $r_3 = r_4$,

so $r_{mNO} = -\dfrac{dN_{NO}}{dt} = 2 L_* L_s k_o \theta_{NO} \theta_H$

Site balance on $*$ sites: $L_* = [*] + [NO*]$ and $K_{NO} = \dfrac{[NO*]}{P_{NO}[*]}$ so

$[*] = \dfrac{L_*}{\left(1 + K_{NO}P_{NO}\right)}$, $[NO*] = \dfrac{L_* K_{NO} P_{NO}}{\left(1 + K_{NO}P_{NO}\right)}$ and $\theta_{NO} = \dfrac{[NO*]}{L_*} = \dfrac{K_{NO}P_{NO}}{1 + K_{NO}P_{NO}}$

Site balance on S sites: $L_s = [S] + [H - S]$ and $K_{H_2} = \dfrac{[H - S]^2}{P_{H_2}[S]^2}$ so

$[S] = \dfrac{L_s}{\left(1 + K_{H_2}^{1/2} P_{H_2}^{1/2}\right)}$, $\theta_H = \dfrac{[H - S]}{L_s} = \dfrac{K_{H_2}^{1/2} P_{H_2}^{1/2}}{1 + K_{H_2}^{1/2} P_{H_2}^{1/2}} = K_{H_2}^{1/2} P_{H_2}^{1/2}$ if $\theta_H << 1$

Then $r_{mNO} = 2 L_* L_s k_o K_{NO} K_{H_2}^{1/2} P_{NO} P_{H_2}^{1/2} / \left(1 + K_{NO}P_{NO}\right)$

(b) $\quad r_{mN_2} = \dfrac{1}{m}\dfrac{dN_{N_2}}{dt} = L_* k_2 \theta_{N_2O}$

SSA on N_2O* gives: $\dfrac{d[N_2O]}{dt} = L_* L_s k_1 \theta_{HNO} \theta_{NO} - L_* k_2 \theta_{N_2O} - L_* k_3 \theta_{N_2O} + L_* k_{-3} P_{N_2O} \theta_* = 0$

so $\left(L_* k_2 + L_* k_3\right)\theta_{N_2O} = L_* L_s k_1 \theta_{HNO} \theta_{NO} + L_* k_{-3} P_{N_2O} \theta_*$ and

$$\theta_{N_2O} = \frac{L_s k_0 \theta_{NO} \theta_H + k_{-3} P_{N_2O} \theta_*}{(k_2 + k_3)} = \frac{L_s k_0 K_{NO} K_{H_2}^{1/2} P_{NO} P_{H_2}^{1/2}}{(k_2 + k_3)(1 + K_{NO} P_{NO})} + \frac{k_{-3} P_{N_2O}}{(k_2 + k_3)(1 + K_{NO} P_{NO})}$$

thus $\quad r_{m_{N_2}} = \dfrac{L_* k_2}{k_2 + k_3} \left[\dfrac{L_s k_0 K_{NO} K_{H_2}^{1/2} P_{NO} P_{H_2}^{1/2} + k_{-3} P_{N_2O}}{(1 + K_{NO} P_{NO})} \right]$

(c) $\quad r_{m_{N_2O}} = \dfrac{1}{m} \dfrac{dN_{N_2O}}{dt} = L_* k_3 \theta_{N_2O} - L_* k_{-3} P_{N_2O} \theta_*$

$$= \frac{L_* k_3}{k_2 + k_3} \left[\frac{L_s k_0 K_{NO} K_{H_2}^{1/2} P_{NO} P_{H_2}^{1/2} + k_{-3} P_{N_2O}}{(1 + K_{NO} P_{NO})} \right] - \frac{L_* k_{-3} P_{N_2O}}{(1 + K_{NO} P_{NO})}$$

$$= \frac{L_*}{k_2 + k_3} \left[\frac{L_s k_3 k_0 K_{NO} K_{H_2}^{1/2} P_{NO} P_{H_2}^{1/2} + k_3 k_{-3} P_{N_2O} - k_2 k_{-3} P_{N_2O} - k_3 k_{-3} P_{N_2O}}{(1 + K_{NO} P_{NO})} \right]$$

$$= \frac{L_*}{k_2 + k_3} \left[\frac{L_s k_0 k_3 K_{NO} K_{H_2}^{1/2} P_{NO} P_{H_2}^{1/2} - k_2 k_{-3} P_{N_2O}}{(1 + K_{NO} P_{NO})} \right]$$

Problem 7.13 Solution

$$d\xi/dt = -\frac{1}{2}\frac{d[N*]}{dt} = k_6[N*]^2$$

$$r = -\frac{d[NH_3]}{dt} = \frac{d[N*]}{dt} = 2k_6[N*]^2 = k_2[NH_3*][*] - k_{-2}[NH_2*][H*]$$

From quasi-equilibrated steps:

$$K_1 = \frac{[NH_3*]}{P_{NH_3}[*]} \quad , \quad K_5 = \frac{P_{H_2}[*]^2}{[H*]^2} \quad , \quad K_3K_4 = \frac{[N*][H*]^2}{[NH_2*][*]^2}$$

Site balance with $[N*]$ as MARI: $L = [*] + [N*]$

SSA on total number of surface N atoms (atomic N_s species):

$$\frac{d[N_s]}{dt} = k_2[NH_3*][*] - k_{-2}[NH_2*][H*] - 2k_6[N*]^2 = 0 \quad \Rightarrow$$

$$K_1k_2P_{NH_3}[*]^2 - \frac{k_{-2}P_{H_2}^{1/2}[*][N*][H*]^2}{K_3K_4K_5^{1/2}[*]^2} - 2k_6[N*]^2 = 0 \quad \Rightarrow$$

$$K_1k_2P_{NH_3}[*]^2 - \frac{k_{-2}P_{H_2}^{3/2}[N*][*]}{K_3K_4K_5^{3/2}} - 2k_6[N*]^2 = 0$$

$$K_1k_2P_{NH_3}(L-[N*])^2 - k_{-2}P_{H_2}^{3/2}\frac{[N*](L-[N*])}{K_3K_4K_5^{3/2}} - 2k_6[N*]^2 = 0$$

$$K_1k_2P_{NH_3}(L^2 - 2L[N*] + [N*]^2) - \frac{Lk_{-2}P_{H_2}^{3/2}[N*]}{K_3K_4K_5^{3/2}} + \frac{k_{-2}P_{H_2}^{3/2}[N*]^2}{K_3K_4K_5^{3/2}} - 2k_6[N*]^2 = 0$$

$$\left(K_1k_2P_{NH_3} + \frac{k_{-2}P_{H_2}^{3/2}}{K_3K_4K_5^{3/2}} - 2k_6\right)[N*]^2 - \left(\frac{2LK_1k_2P_{NH_3} + Lk_{-2}P_{H_2}^{3/2}}{K_3K_4K_5^{3/2}}\right)[N*] + L^2K_1k_2P_{NH_3} = 0$$

Solution for quadratic equation: $ax^2 + bx + c = 0$ is $\dfrac{-b \pm \sqrt{b^2 - 4ac}}{2a}$ so

$$[N*] = \frac{\left(2LK_1k_2P_{NH_3} + \dfrac{Lk_{-2}P_{H_2}^{3/2}}{K_3K_4K_5^{3/2}}\right)}{2\left(K_1k_2P_{NH_3} + \dfrac{k_{-2}P_{H_2}^{3/2}}{K_3K_4K_5^{3/2}} - 2k_6\right)} \pm$$

$$\frac{\left[\left(\dfrac{2LK_1k_2P_{NH_3} + Lk_{-2}P_{H_2}^{3/2}}{K_3K_4K_5^{3/2}}\right)^2 - 4L^2K_1k_2P_{NH_3}\left(K_1k_2P_{NH_3} + \dfrac{k_{-2}P_{H_2}^{3/2}}{K_3K_4K_5^{3/2}} - 2k_6\right)\right]^{1/2}}{2\left(K_1k_2P_{NH_3} + \dfrac{k_{-2}P_{H_2}^{3/2}}{K_3K_4K_5^{3/2}} - 2k_6\right)}$$

Divide numerator and denominator by -2 k_6 and use positive root:

$$[N*] = \frac{-\left(\dfrac{2LK_1k_2P_{NH_3}}{k_6} + \dfrac{Lk_{-2}P_{H_2}^{3/2}}{K_3K_4K_5^{3/2}k_6}\right)}{4\left(1 - \dfrac{K_1k_2P_{NH_3}}{k_6} - \dfrac{k_{-2}P_{H_2}^{3/2}}{K_3K_4K_5^{3/2}k_6}\right)} +$$

$$\frac{\left[\left(\dfrac{2LK_1k_2P_{NH_3}}{k_6} + \dfrac{Lk_{-2}P_{H_2}^{3/2}}{K_3K_4K_5^{3/2}k_6}\right)^2 - \dfrac{4L^2K_1k_2P_{NH_3}}{k_6}\left(K_1k_2P_{NH_3} + \dfrac{k_{-2}P_{H_2}^{3/2}}{K_3K_4K_5^{3/2}} - 2k_6\right)\right]^{1/2}}{4\left(1 - \dfrac{K_1k_2P_{NH_3}}{k_6} - \dfrac{k_{-2}P_{H_2}^{3/2}}{K_3K_4K_5^{3/2}k_6}\right)}$$

Inside [----]$^{1/2}$ after factoring L out:

$$\frac{4K_1^2k_2^2P_{NH_3}^2}{k_6^2} + \frac{4K_1k_2k_{-2}P_{NH_3}P_{H_2}^{3/2}}{K_3K_4K_5^{3/2}k_6} + \frac{k_{-2}^2P_{H_2}^3}{K_3^2K_4^2K_5^3k_6^2} - \frac{4K_1^2k_2^2P_{NH_3}^2}{k_6^2} - \frac{4K_1k_2k_{-2}P_{NH_3}P_{H_2}^{3/2}}{K_3K_4K_5^{3/2}k_6} + \frac{8K_1k_2P_{NH_3}}{k_6}$$

so

$$[N*] = \dfrac{\dfrac{L}{4}\left(\dfrac{-2K_1k_2P_{NH_3}}{k_6} - \dfrac{k_{-2}P_{H_2}^{3/2}}{K_3K_4K_5^{3/2}k_6}\right) + \left[\dfrac{8K_1k_2P_{NH_3}}{k_6} + \dfrac{k_{-2}^2P_{H_2}^3}{K_3^2K_4^2K_5^3k_6^2}\right]^{1/2}}{\left(1 + \dfrac{K_1k_2P_{NH_3}}{k_6} - \dfrac{k_{-2}P_{H_2}^{3/2}}{K_3K_4K_5^{3/2}k_6}\right)}$$

and, after introducing probability factor $Z/2L$ for neighbor sites:

$$r = \dfrac{ZLk_6}{4}\left[\dfrac{-2aP_{NH_3} - bP_{H_2}^{3/2} + \left(8aP_{NH_3} + b^2P_{H_2}^3\right)^{1/2}}{\left(1 - aP_{NH_3} - bP_{H_2}^{3/2}\right)}\right]^2 \qquad \text{where}$$

$$a = \dfrac{K_1k_2}{k_6} \text{ and } b = \dfrac{k_{-2}}{K_3K_4K_5^{3/2}k_6}$$

Problem 7.14 Solution

The specific rate is:
$$r_m = -\frac{1}{m}\frac{dN_{N_2O}}{dt} = \frac{1}{m}\frac{dN_{N_2}}{dt} = Lk_1\theta_{N_2O}$$

From steps 1 and 2:
$$K_{N_2O} = \frac{\theta_{N_2O}}{P_{N_2O}\theta_v} \text{ and } K_{CO} = \frac{\theta_{CO}}{P_{CO}\theta_v}$$

SSA on O* :
$$\frac{d\theta_O}{dt} = Lk_1\theta_{N_2O} - Lk_2\theta_{CO}\theta_O = 0 \quad \Rightarrow$$

$$\theta_O = \frac{k_1\theta_{N_2O}}{k_2\theta_{CO}} = \frac{k_1 K_{N_2O}P_{N_2O}\theta_v}{k_2 K_{CO}P_{CO}\theta_v}$$

Site balance:
$$1 = \theta_v + K_{N_2O}P_{N_2O}\theta_v + K_{CO}P_{CO}\theta_v + \frac{k_1 K_{N_2O}P_{N_2O}}{k_2 K_{CO}P_{CO}}$$

$$\left(1 - \frac{k_1 K_{N_2O}P_{N_2O}}{k_2 K_{CO}P_{CO}}\right) = \theta_v\left(1 + K_{N_2O}P_{N_2O} + K_{CO}P_{CO}\right)$$

$$r_m = Lk_1 K_{N_2O}P_{N_2O}\theta_v = \frac{Lk_1 K_{N_2O}P_{N_2O}\left(1 - \dfrac{k_1 K_{N_2O}P_{N_2O}}{k_2 K_{CO}P_{CO}}\right)}{\left(1 + K_{N_2O}P_{N_2O} + K_{CO}P_{CO}\right)}$$

or
$$r_m = \left(\mu\text{mole s}^{-1}\text{ g}^{-1}\right) = \frac{cP_{N_2O} - dP_{N_2O}^2/P_{CO}}{\left(1 + K_{N_2O}P_{N_2O} + K_{CO}P_{CO}\right)}$$

where $c = Lk_1 K_{N_2O}$ and $d = k_1^2 K_{N_2O}^2/k_2 K_{CO}$

An Arrhenius plot of Lk_1 gives $E_1 = 26$ kcal mole^{-1} while similar plots for K_{N_2O} and K_{CO} give:

for N_2O: $\Delta H_{ad}^o = -13$ kcal mole^{-1} and $\Delta S_{ad}^o = -29$ cal mole^{-1} K^{-1} (e.u.)

for CO: $\Delta H_{ad}^o = -10$ kcal mole^{-1} and $\Delta S_{ad}^o = -23$ e.u.

All these thermodynamic values satisfy the criteria in Table 6.9. The TOF at 573K =

$$\frac{2100\,\mu\text{mole N2 g}^{-1}\,\text{s}^{-1}}{(769\,\mu\text{mole Cu g}^{-1})(0.91)} = 3.0\,\text{s}^{-1}.$$ Thus the pre-exponential factor per site is: $3.0\,\text{s}^{-1} =$

$A_o e^{-26000/1.987\,(573)}$ and $A_o = 2.4 \times 10^{10}\,\text{s}^{-1}$ which is less than $10^{13}\,\text{s}^{-1}$ and satisfies criterion 2

in Table 6.10 for a unimolecular reaction. Also, the heat of adsorption of 10 kcal mole^{-1}

for CO on Cu is consistent with values of 7-10 kcal mole^{-1} reported in the literature.

Problem 7.15 Solution

$$r = -\frac{d[CH_4]}{dt} = k_1 P_{CH_4}[*] - k_{-1}[CH_2 *]P_{H_2} = k_7[CH_2O *]$$

from step (6): $[CH_2O *] = K_6[CH_2 *][OH *]/[H *]$

from step (4): $[OH *] = K_4[CO_2 *][H *]/[CO *]$

from step (2): $[CO_2 *] = K_2 P_{CO_2}[*]$

from step (8): $[CO *] = K_8 P_{CO}[*]$ therefore

$$r = k_7 K_6[CH_2 *][OH *]/[H *] = k_7 K_6 K_4[CH_2 *][CO_2 *]/[CO *]$$

$$= k_7 K_6 K_4 K_2 P_{CO_2}[CH_2 *]/K_8 P_{CO}$$

SSA on surface $CH_2 *$ (and $CH_2O *$) species:

$$\frac{d[CH_2 \cdots *]}{dt} = k_1 P_{CH_4}[*] - k_{-1}[CH_2 *]P_{H_2} - k_7[CH_2O *] = 0 \qquad \Rightarrow$$

$$[CH_2O *] = \frac{(k_1 P_{CH_4}[*] - k_{-1}[CH_2 *]P_{H_2})}{k_7}$$

from step (6):
$$[CH_2*] = \frac{[CH_2O*][H*]}{K_6[OH*]} = \frac{[CH_2O*][CO*]}{K_6K_4[CO_2*]} =$$

$$\frac{K_8[CH_2O*]P_{CO}[*]}{K_2K_4K_6P_{CO_2}[*]} = \left(\frac{K_8P_{CO}}{K_2K_4K_6P_{CO_2}}\right)[CH_2O*]$$

from SSA: $\quad k_1P_{CH_4}[*] = \left(k_7 + \dfrac{k_{-1}K_8P_{CO}P_{H_2}}{K_2K_4K_6P_{CO_2}}\right)[CH_2O*]$ \quad so

$$[CH_2O*] = \frac{(k_1/k_7)P_{CH_4}[*]}{\left(1 + \dfrac{k_{-1}K_8P_{CO}P_{H_2}}{k_7K_2K_4K_6P_{CO}}\right)}$$

Site balance with $[CH_2O*]$ as MARI: $\quad L = [*] + [CH_2O*] =$

$$[*]\left[1 + \frac{(k_1P_{CH_4}/k_7)}{\left(1 + k_{-1}K_8P_{CO}P_{H_2}/k_7K_2K_4K_6P_{CO_2}\right)}\right] =$$

$$[*]\left[\frac{1 + \dfrac{k_{-1}K_8P_{CO}P_{H_2}}{k_7K_2K_4K_6P_{CO_2}} + \dfrac{k_1P_{CH_4}}{k_7}}{1 + \dfrac{k_{-1}K_8P_{CO}P_{H_2}}{k_7K_2K_4K_6P_{CO_2}}}\right] \quad so$$

$$[*] = \frac{L\left(1 + \dfrac{k_{-1}K_8P_{CO}P_{H_2}}{k_7K_2K_4K_6P_{CO_2}}\right)}{\left(1 + \dfrac{k_1P_{CH_4}}{k_7} + \dfrac{k_{-1}K_8P_{CO}P_{H_2}}{k_7K_2K_4K_6P_{CO_2}}\right)}$$

$$r = k_7[CH_2O*] = \frac{k_7(k_1/k_7)P_{CH_4}[*]}{\left(1 + \dfrac{k_{-1}K_8P_{CO}P_{H_2}}{k_7K_2K_4K_6P_{CO}}\right)} = \frac{Lk_1P_{CH_4}}{\left(1 + \dfrac{k_1P_{CH_4}}{k_7} + \dfrac{k_{-1}K_8P_{CO}P_{H_2}}{k_7K_2K_4K_6P_{CO_2}}\right)}$$

or $$r = \frac{Lk_1P_{CH_4}P_{CO_2}}{\left(P_{CO_2} + \dfrac{k_1}{k_7}P_{CH_4}P_{CO_2} + \dfrac{k_{-1}K_8P_{CO}P_{H_2}}{k_7K_2K_4K_6}\right)}$$

The TOF for step 7 is: $\dfrac{\left(5.35\ \mu\text{mole s}^{-1}\ \text{g}^{-1}\right)}{\left(2.4\ \mu\text{mole H}_2\ \text{g}^{-1}\right)\left(\dfrac{2\ \mu\text{mole Ni}_s}{\mu\text{mole H}_2}\right)} = 1.1\ \text{s}^{-1}$

Rate per site: TOF = 1.1 s^{-1} = $Ae^{-E/RT}$ = $Ae^{-38000/1.987\,(723)}$ = A (3.25 x 10^{-12}). Therefore A = 1.1/3.25 x 10^{-12} = 3.4 x 10^{11} s^{-1}, which is less than 10^{13} (See Criterion 2 in Table 6.10)

Problem 7.16 Solution

(a) $\quad r_m = \dfrac{1}{m}\dfrac{d\xi}{dt} = \dfrac{1}{m}\dfrac{dN_i}{v_i dt} = \dfrac{-1}{2m}\dfrac{dN_{NO}}{dt} = \dfrac{1}{m}\dfrac{dN_{N_2}}{dt} \Rightarrow \dfrac{dN_2}{dt} = \dfrac{-1}{2}\dfrac{dN_{NO}}{dt}$

$-\dfrac{dN_{NO}}{dt} = 2k_1[NO*]^2 \quad , \quad K_{NO} = \dfrac{[NO*]}{P_{NO}[*]} \Rightarrow [NO*] = K_{NO}P_{NO}[*]$

$L = [*] + 2[*(NO)_2 *] + [N_2O*] + [O*] + [NO*] = [*] + [NO*] + [O*]$

$K_{O_2} = \dfrac{[O*]^2}{P_{O_2}[*]^2} \quad \Rightarrow \quad [O*] = K_{O_2}^{1/2}P_{O_2}^{1/2}[*]$

$L = [*]\left(1 + K_{NO}P_{NO} + K_{O_2}^{1/2}P_{O_2}^{1/2}\right)$ and $[*] = L/\left(1 + K_{NO}P_{NO} + K_{O_2}^{1/2}P_{O_2}^{1/2}\right)$

So $\quad r_{N_2} = \dfrac{2k_1}{2}[NO*]^2 = k_1 K_{NO}^2 P_{NO}^2[*]^2 = \dfrac{L k_1 K_{NO}^2 P_{NO}^2}{\left(1 + K_{NO}P_{NO} + K_{O_2}^{1/2}P_{O_2}^{1/2}\right)^2}$

or

$r_{N_2} = k_3[N_2O*]$ and SSA on N_2O* gives $k_2[*(NO)_2 *] = k_3[N_2O*]$

so $\quad r_{N_2} = k_2[*(NO)_2 *]$ and SSA on $[*(NO)_2 *]$ gives $k_1[NO*]^2 = k_2[*(NO)_2 *]$

so $\quad r_{N_2} = k_1[NO*]^2 = Lk_1 K_{NO}^2 P_{NO}^2 / \left(1 + K_{NO}P_{NO} + K_{O_2}^{1/2}P_{O_2}^{1/2}\right)^2$

(b) In the site balance:

$k_3[N_2O*] = k_1[NO*]^2$ and $[N_2O*] = (k_1/k_3)K_{NO}P_{NO}[*]^2$

and $k_2[*(NO)_2 *] = k_1[NO*]^2$ and $[*(NO)_2 *] = (k_1/k_2)K_{NO}P_{NO}[*]^2$

so $\quad L = [*] + \dfrac{2k_1 K_{NO}P_{NO}}{k_2}[*]^2 + \dfrac{k_1 K_{NO}P_{NO}}{k_3}[*]^2 + K_{NO}P_{NO}[*] + K_{O_2}^{1/2}P_{O_2}^{1/2}[*]$

Consequently, a quadratic expression for $[*]$ is obtained and the solution for such an equation is complicated.

Problem 7.17 Solution

The series of elementary steps is:

(1) $\quad MCH + * \underset{}{\overset{K_1}{\rightleftharpoons}} MCH*$

(2) $\quad MCH* \underset{}{\overset{K_2}{\rightleftharpoons}} MCX* + H_2$

(3) $\quad MCX* \underset{}{\overset{K_3}{\rightleftharpoons}} MCD* + H_2$

(4) $\quad MCD* \underset{}{\overset{K_4}{\rightleftharpoons}} TOL* + H_2$

(5) $\quad \underline{\quad TOL* \overset{k_5}{\longrightarrow} TOL + * \quad}$ (RDS)

$\quad\quad MCH \implies TOL$

or

$MCH + * \underset{}{\overset{K'}{\rightleftharpoons}} TOL* + 3H_2 \quad\quad$ where $\ K' = K_1 K_2 K_3 K_4$

$TOL* \overset{k_5}{\longrightarrow} TOL + *$

Step 5 is RDS, so $r = k_5 [TOL*]$

$$K' = \frac{[TOL*]P_{H_2}^3}{P_{MCH}[*]} \quad\quad \text{and} \quad\quad [TOL*] = \frac{K'P_{MCH}[*]}{P_{H_2}^3}$$

$$L = [*] + [MCH*] + [MCX*] + [MCD*] + [TOL*]$$

If $[TOL*]$ is MARI, then $L = [*] + [TOL*] = [*]\left(1 + K'P_{MCH}/P_{H_2}^3\right)$ and $[*] = \dfrac{L}{\left(1 + K'P_{MCH}/P_{H_2}^3\right)}$

then $\quad r = k_5[TOL*] = k_5 K'P_{MCH}[*]/P_{H_2}^3 = \dfrac{Lk_5 K'P_{MCH}}{\left(1 + K'P_{MCH}/P_{H_2}^3\right)P_{H_2}^3}$

or $\quad r = \dfrac{Lk_5 K'P_{MCH}}{\left(P_{H_2}^3 + K'P_{MCH}\right)}$

This is not consistent with the behavior because it contains a $P_{H_2}^3$ term.

Problem 7.18 Solution

$$\frac{-d[O_{2(g)}]}{dt} = r = 2k_1 P_{O_2}[S]^2 = k_3[HCO_2 - S][O - S] = kP_{O_2}[S]^2 \qquad L = [S] + [HCO_2 - S] + [O - S]$$

A. $\quad [HCO_2 - S] = K_2 P_{H_2CO}[O - S]^2 / [OH - S]$

B. $\quad [OH - S] = K_{H_2O}^{1/2} P_{H_2O}^{1/2}[O - S]^{1/2}[S]^{1/2}$

C. $\quad [HCO_2 - S] = \dfrac{K_2 P_{H_2CO}[O - S]^2}{K_{H_2O}^{1/2} P_{H_2O}^{1/2}[O - S]^{1/2}[S]^{1/2}} =$

$$\frac{K_2 P_{H_2CO}[O - S]^{3/2}[S]^{-1/2}}{K_{H_2O}^{1/2} P_{H_2O}^{1/2}}$$

D. $\quad$ At steady-state: $\quad [O - S] = \dfrac{k_1 P_{O_2}[S]^2}{k_3[HCO_2 - S]} \qquad$ (Balance on O-S species)

E. $\quad$ Substitute D in C: $\quad [HCO_2 - S] = \dfrac{K_2 P_{H_2CO} k_1^{3/2} P_{O_2}^{3/2}[S]^{5/2}}{k_3^{3/2}[HCO_2 - S]^{3/2} K_{H_2O}^{1/2} P_{H_2O}^{1/2}} \qquad \Rightarrow$

F. $\quad [HCO_2 - S] = \dfrac{K_2^{2/5} P_{H_2CO}^{2/5} k_1^{3/5} P_{O_2}^{3/5}[S]}{k_3^{3/5} K_{H_2O}^{1/5} P_{H_2O}^{1/5}}$

G. $\quad [O - S] = \dfrac{k_1 P_{O_2}^{2/5}[S] K_{H_2O}^{1/5} P_{H_2O}^{1/5}}{k_3^{2/5} K_2^{2/5} k_1^{3/5} P_{H_2CO}^{2/5}}$

H. $\quad L = [S] + \dfrac{K_2^{2/5} k_1^{3/5} P_{H_2CO}^{2/5} P_{O_2}^{3/5}[S]}{k_3^{3/5} K_{H_2O}^{1/5} P_{H_2O}^{1/5}} + \dfrac{k_1^{2/5} P_{O_2}^{2/5} K_{H_2O}^{1/5} P_{H_2O}^{1/5}[S]}{k_3^{2/5} K_2^{2/5} P_{H_2CO}^{2/5}}$

I. $\quad [S] = L / \left(1 + K'P_{H_2CO}^{2/5} P_{O_2}^{3/5} P_{H_2O}^{-1/5} + K''P_{O_2}^{2/5} P_{H_2CO}^{-2/5} P_{H_2O}^{1/5}\right)^2 \qquad$ Assume P_{H_2O} is $\approx$ constant, plus power of 0.2 is low

Then, $\quad r = Lk_1 P_{O_2} / \left(1 + K'''P_{H_2CO}^{0.4} P_{O_2}^{0.6} + K''''P_{O_2}^{0.4} P_{H_2CO}^{-0.4}\right)^2$

If $\theta_{HCO_2} \ll 1$ and $\theta_O \ll 1$, then $r = kP_{O_2}$

If θ_O is MASI, i.e., $\theta_O \gg \theta_{HCO_2}$ then

$$r = \frac{Lk_1 P_{O_2}}{\left(1 + K''' P_{O_2}^{0.4}/P_{H_2CO}^{0.4}\right)^2} = \frac{k' P_{O_2} P_{H_2CO}^{0.8}}{\left(P_{H_2CO}^{0.4} + K''' P_{O_2}^{0.4}\right)^2}$$

(Note typographical error in power dependence on H_2O in ref. 62)

Problem 8.1 Solution

$$r = \frac{k_1 k_2 [A_1][A_2] - k_{-1} k_{-2}[B_1][B_2]}{k_1[A_1] + k_{-1}[B_1] + k_2[A_2] + k_{-2}[B_2]} \qquad \alpha = 2/3$$

$$k_1 = k_1^o e^{\alpha(t-t_o)} = k_1^o e^{2/3(t-t_o)} \;,\; k_{-1} = k_{-1}^o e^{-1/3(t-t_o)} \;,\; k_2 = k_2^o e^{-1/3(t-t_o)} \;,\; k_{-2} = k_{-2}^o e^{2/3(t-t_o)}$$

Substitute into r:

$$r = \frac{k_1^o k_2^o e^{2/3(t-t_o)} e^{-1/3(t-t_o)}[A_1][A_2] - k_{-1}^o k_{-2}^o e^{-1/3(t-t_o)} e^{2/3(t-t_o)}[B_1][B_2]}{k_1^o e^{2/3(t-t_o)}[A_1] + k_{-1}^o e^{-2/3(t-t_o)}[B_1] + k_2^o e^{-2/3(t-t_o)}[A_2] + k_{-2}^o e^{1/3(t-t_o)}[B_2]}$$

$$-k_{-1}^o k_{-2}^o$$

$$r = \frac{k_1^o k_2^o [A_1][A_2] - k_{-1}^o k_{-2}^o [B_1][B_2]}{\left[E_1 e^{1/3(t-t_o)} + E_2 e^{-2/3(t-t_o)}\right]} \qquad \begin{array}{l} \text{where } E_1 = k_1^o[A_1] + k_{-2}^o[B_2] \\ \text{and} \quad E_2 = k_{-1}^o[B_1] + k_2^o[A_2] \end{array}$$

Let D represent the denominator, then

$$dD/dt = 0 = 1/3 E_1 e^{-1/3(t_{max}-t_o)} - 2/3 E_2 e^{-2/3(t_{max}-t_o)} \qquad \Rightarrow$$

$$1/3 E_1 e^{1/3(t_{max}-t_o)} = 2/3 E_2 e^{-2/3(t_{max}-t_o)} \qquad \Rightarrow$$

$$E_2 \big/ E_1 = 1/2 e^{(t_{max}-t_o)} \;\;; \;\; u_{max} = u_o e^{(t_o-t_{max})} = \frac{[S_1]}{[S_2]}$$

$$u_o = E_2/E_1 \qquad \Rightarrow \qquad u_{max} = \left[1/2 e^{(t_{max}-t_o)}\right] e^{(t_o-t_{max})} = 1/2$$

$$\frac{[S_1]}{[S_2]} = 1/2 \;\; \Rightarrow \;\; 2[S_1] = [S_2] \;\; \text{ and } \;\; \theta = \frac{[S_2]}{[S_1]+[S_2]} = \frac{[S_2]}{1/2[S_2]+[S_2]} = \frac{2}{3}$$

Problem 8.2 Solution

Because $k_1^o P_{N_2} \ll k_{-2}^o P_{NH_3}^2$ and $k_{-1}^o \ll k_2^o P_{H_2}^3$:

$$r = k \left[\frac{k_1^o k_2^o [N_2][H_2]^3 - k_{-1}^o k_{-2}^o [NH_3]^2}{k_{-2}^{o^{\,m}}[NH_3]^{2m} k_2^{o^{(1-m)}}[H_2]^{3(1-m)}} \right] \qquad , \qquad \text{thus}$$

$$r = k \left[\frac{k_1^o k_2^{o^{\,m}} P_{N_2} P_{H_2}^{3m}}{k_{-2}^m P_{NH_3}^{2m}} - \frac{k_{-1}^o k_{-2}^{o^{(1-m)}} P_{NH_3}^{2(1-m)}}{k_2^{o^{(1-m)}} P_{H_2}^{3(1-m)}} \right] =$$

$$k \left[k_1^o K_2^{o^{\,m}} \frac{P_{N_2} P_{H_2}^{3m}}{P_{NH_3}^{2m}} - \frac{k_{-1}^o P_{NH_3}^{2(1-m)}}{K_2^{o^{(1-m)}} P_{H_2}^{3(1-m)}} \right] \qquad , \qquad m = 1/2 \quad \text{so}$$

$$r = k \left[k_1^o K_2^{o^{1/2}} \frac{P_{N_2} P_{H_2}^{3/2}}{P_{NH_3}} - \frac{k_{-1}^o P_{NH_3}}{K_2^{o^{1/2}} P_{H_2}^{3/2}} \right] = \bar{k} \frac{P_{N_2} P_{H_2}^{3/2}}{P_{NH_3}} - \bar{k} \frac{P_{NH_3}}{P_{H_2}^{3/2}}$$

Problem 9.1 Solution

$$A + E \; \overset{k_1}{\underset{k_{-1}}{\rightleftharpoons}} \; A \cdot E$$

$$A \cdot E \; \overset{k_2}{\underset{k_{-2}}{\rightleftharpoons}} \; P + E$$

$$A \implies P$$

$$r = \frac{d[P]}{dt} = \vec{r} - \overleftarrow{r} = k_2 [A \cdot E] - k_{-2} [P][E]$$

SSA on $A \cdot E$: $\; k_1 [A][E] + k_{-2} [P][E] - k_{-1} [A \cdot E] - k_2 [A \cdot E] = 0$

Active site balance: $L_e = [E] + [A \cdot E]$

$$[A \cdot E](k_{-1} + k_2) = k_1 [A][E] = k_{-2} [P][E] \qquad \text{from SSA} \; \Rightarrow$$

$$[A \cdot E] = \frac{k_1 [A][E] + k_{-2} [P][E]}{k_{-1} + k_2}$$

$$L_e = [E] + \frac{k_1 [A][E] + k_{-2} [P][E]}{k_{-1} + k_2} = \frac{(k_{-1} + k_2)[E] + k_1 [A][E] + k_{-2} [P][E]}{k_{-1} + k_2}$$

$$r = k_2 \left(\frac{k_1 [A][E] + k_{-2} [P][E]}{k_{-1} + k_2} \right) - k_{-2} [P][E] = \left(\frac{k_1 k_2 [A] + k_2 k_{-2} [P] - k_{-1} k_{-2} [P] - k_2 k_{-2} [P]}{k_{-1} + k_2} \right)[E]$$

$$r = \frac{(k_{-1} + k_2) L_e}{(k_{-1} + k_2) + k_1 [A] + k_{-2} [P]} \cdot \frac{(k_1 k_2 [A] - k_{-1} k_{-2} [P])}{(k_{-1} + k_2)} = \frac{L_e (k_1 k_2 [A] - k_{-1} k_{-2} [P])}{(k_{-1} + k_2) + k_1 [A] + k_{-2} [P]}$$

Initial forward rate, $[P] \cong 0 \Rightarrow \dfrac{L_e k_2 [A]}{\left(\dfrac{k_{-2}+k_2}{k_1}\right)+[A]} = r \Rightarrow$ same form as eq. 9.6

Initial reverse rate, $[A] \cong 0$, $r = -\dfrac{d[P]}{dt} = \dfrac{d[A]}{dt} = \dfrac{L_e k_{-1}[P]}{\left(\dfrac{k_{-1}+k_2}{k_{-2}}\right)+[P]} \Rightarrow$

also of Michaelis-Menten form.

Problem 9.2 Solution

Use eq. 9.9, which is: $1/r = \left(\dfrac{K_m}{r_{max}}\right)\left(\dfrac{1}{[A]}\right) + \dfrac{1}{r_{max}}$ and plot $\dfrac{1}{r}$ vs. $\dfrac{1}{[A]}$

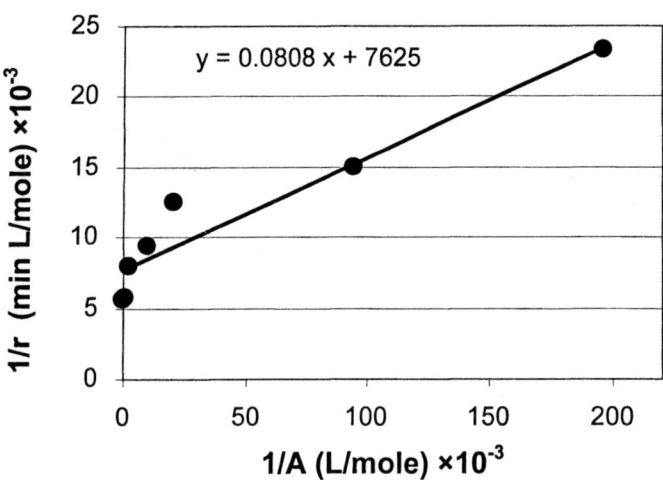

From the slope and the intercept:

$$\text{Intercept} \;=\; \dfrac{1}{r_{max}} = 7620 \;\; so \; r_{max} = 1.31 \times 10^{-4} \; \dfrac{mole}{L \cdot min}$$

$$\text{Slope} \;= 0.0808 \;\; min = \dfrac{K_m}{r_{max}} \;\;\Rightarrow$$

$$K_m = (0.0808 \;\; min)\left(1.31 \times 10^{-4} \; mole/L \cdot min\right) = 1.06 \times 10^{-5} \; \dfrac{mole}{L}$$

Problem 9.3 Solution

$$r = \frac{d[P]}{dt} = k[ES]$$

Steps 1 and 2 are quasi-equilibrated, so

$$K_s = \frac{[ES]}{[E][S]} \quad \text{and} \quad K'_s = \frac{[ESS]}{[ES][S]}$$

Balance on enzyme: $L_e = [E] + [ES] + [ESS]$

$$L_e = \frac{[ES]}{K_s[S]} + [ES] + K'_s[ES][S], \text{ and } [ES] = \frac{L_e}{1 + \dfrac{1}{K_s[S]} + K'_s[S]}$$

$$r = \frac{L_e k}{\left(1 + \dfrac{1}{K_s[S]} + K'_s[S]\right)} = \frac{L_e k[S]}{\left(1/K_s + [S] + K'_s[S]^2\right)}$$

Note: by convention, steps 1 and 2 are represented by a dissociation equilibrium

constant, i.e.,

$$ES \; \underset{K_1}{\overset{}{\rightleftharpoons}} \; E + S$$

$$ESS \; \underset{K_2}{\overset{}{\rightleftharpoons}} \; ES + S$$

so then $K_1 = 1/K_s$ and $K_2 = 1/K'_s$ and $r = \dfrac{L_e k[S]}{\left(K_1 + [S] + [S]^2/K_2\right)}$

written in terms of dissociation constants. The mathematical form is identical to the

above rate equation.

Problem 9.4 Solution

(a)   Steps 1 and 2 are in quasi-equilibrium:

(1) $S + E \xrightleftharpoons[k_{-1}]{k_1} (ES)_1$ $K_1 = \dfrac{[(ES)_1]}{[S][E]} = \dfrac{k_1}{k_{-1}}$

(2) $(ES)_1 \xrightleftharpoons[k_{-2}]{k_2} (ES)_2$ $K_2 = \dfrac{[(ES)_2]}{[(ES)_1]} = \dfrac{k_2}{k_{-2}}$

(3) $(ES)_2 \xrightarrow{\ k\ } P + E$

$S \longrightarrow P$

$$r = \frac{d[P]}{dt} = k[(ES)_2] = kK_2[(ES)_1] = kK_1K_2[S][E]$$

Active site balance: $L_e = E + (ES)_1 + (ES)_2$

$$L_e = [E] + K_1[S][E] + K_1K_2[S][E] = [E](1 + K_1[S] + K_1K_2[S])$$

$$r = \frac{L_e kK_1K_2[S]}{1 + (K_1 + K_1K_2)[S]} = \frac{L_e kK_2[S]}{1/K_1 + (1 + K_2)[S]} = \frac{\left(\dfrac{L_e kK_2}{1 + K_2}\right)[S]}{\left(\dfrac{1}{K_1 + K_1K_2} + [S]\right)}$$

so r_{max} (apparent) $= L_e kK_2/(1 + k_2)$ and K_m (apparent) $= 1/(K_1 + K_1K_2)$

(b) Steps 1 and 2 are reversible, so use SSA:

(1) $\dfrac{d[(ES)_1]}{dt} = k_1[S][E] + k_{-2}[(ES)_2] - k_{-1}[(ES)_1] - k_2[(ES)_1] = 0$

(2) $\dfrac{d[(ES)_2]}{dt} = k_2[(ES)_1] - k_{-2}[(ES)_2] - k[(ES)_2] = 0$

(3) Enzyme balance: $L_e = [E] + [(ES)_1] + [(ES)_2]$

From (2): $[(ES)_2](k_{-2} + k) = k_2[(ES)_1] \Rightarrow [(ES)_2] = k_2[(ES)_1]/(k_{-2} + k)$

From (1): $k_1[S][E] = .\dfrac{k_2 k_{-2}[(ES)_1]}{k_{-2} + k}. - (k_{-1} + k_2)[(ES)_1] = 0 \Rightarrow$

$k_1[S][E] = \left(k_{-1} + k_2 - \dfrac{k_2 k_{-2}}{k_{-2} + k}\right)[(ES)_1] \Rightarrow$

$[(ES)_1] = \dfrac{k_1[S][E]}{k_{-1} + k_2 - \dfrac{k_2 k_{-2}}{k_{-2} + k}} = \dfrac{k_1[S][E](k_{-2} + k)}{k_{-1}k_{-2} + k_2 k_{-2} + k_{-1}k + k_2 k - k_2 k_{-2}} = \dfrac{k_1(k_{-2} + k)[S][E]}{k_{-1}(k_{-2} + k) + k_2 k}$

$L_e = [E] + \dfrac{k_1[S][E]}{k_{-1} + k_2 - \dfrac{k_2 k_{-2}}{k_{-2} + k}} + \dfrac{k_1 k_2[S][E]/(k_{-2} + k)}{k_{-1} + k_2 - \dfrac{k_2 k_{-2}}{k_{-2} + k}} = [E] + \dfrac{(k_1(k_{-2} + k) + k_1 k_2)[S][E]}{k_{-1}(k_{-2} + k) + k_2 k}$

$L_e = [E]\left[\dfrac{1 + (k_1(k_{-2} + k) + k_1 k_2)[S]}{k_{-1}(k_{-2} + k) + k_2 k}\right] = [E]\left[\dfrac{k_{-1}(k_{-2} + k) + k_2 k + k_1(k_2 + k_{-2} + k)[S]}{k_{-1}(k_{-2} + k) + k_2 k}\right]$

$[E] = \dfrac{L_e[k_{-1}(k_2 + k) + k_2 k]}{k_{-1}(k_{-2} + k) + k_2 k + k_1(k_2 + k_{-2} + k)[S]}$

$$r = k[(ES)_2] = \frac{kk_2[(ES)_1]}{(k_{-2}+k)} = \frac{kk_1k_2[S][E]}{[k_{-1}(k_{-2}+k)+k_2k]} = \frac{L_ekk_1k_2[S]}{k_{-1}(k_{-2}+k)+k_2k+k_1(k_2+k_{-2}+k)[S]}$$

To get Michaelis-Menten form, divide numerator and denominator by

$$k_1(k_2+k_{-2}+k) \quad \Rightarrow$$

$$r = \frac{\left(\dfrac{L_ekk_2}{k_2+k_{-2}+k}\right)[S]}{\dfrac{k_{-1}k_{-2}+k_{-1}k+k_2k}{k_1(k_2+k_{-2}+k)}+[S]}$$

so r_{max} (apparent) $= \dfrac{L_ekk_2}{k_2+k_{-2}+k}$

and K_m (apparent)) $= \dfrac{k_{-1}k_{-2}+k_{-1}k+k_2k}{k_1k_2+k_1k_{-2}+k_1k}$

Problem 9.5 Solution

$$[E] \overset{K_1}{\rightleftharpoons} [E^-] + [H^+] \qquad\qquad K_1 = \frac{[E^-][H^+]}{E}$$

$$[E^-] \overset{K_2}{\rightleftharpoons} [E^{-2}] + [H^+] \qquad\qquad K_2 = \frac{[E^{-2}][H^+]}{[E^-]}$$

$$L_e = [E] + [E^-] + [E^{-2}]$$

$$\text{active fraction} = y^- = \frac{[E^-]}{L_e} = \frac{[E^-]}{\left([E]+[E^-]+[E^{-2}]\right)}$$

$$y^- = \frac{[E^-]}{\left(\dfrac{[H^+][E^-]}{K_1}+[E^-]+K_2[E^-]\Big/[H^+]\right)} = \frac{1}{\left(1+[H^+]/K_1+K_2/[H^+]\right)} \quad \text{while}$$

$$y = \frac{[E]}{L_e} = \frac{1}{\left(1+K_1/[H^+]+K_1K_2/[H^+]^2\right)} \quad \text{and}$$

$$y^{-2} = \frac{[E^{-2}]}{L_e} = \frac{1}{\left(1+[H^+]/K_2+[H^+]^2/K_1K_2\right)}$$

Thus the maximum rate is proportional to the active fraction and

$$r_{max} = L_e k_2 y^- = \frac{L_e k_2}{\left(1+[H^+]/K_1+K_2/[H^+]\right)}$$